Planning Land 3-D Seismic Surveys

Andreas Cordsen, Mike Galbraith,
and John Peirce

Edited by Bob A. Hardage
Series Editor: Stephen J. Hill

Geophysical Developments Series No. 9
Society of Exploration Geophysicists

SOCIETY OF EXPLORATION GEOPHYSICISTS

Library of Congress Cataloging-in-Publication Data

Cordsen, Andreas.
    Planning land 3-D seismic surveys / by Andreas Cordsen, Mike Galbraith, and John Peirce ; editor, Bob A. Hardage.
       p. cm. — (Geophysical developments series ; no. 9)
    Includes bibliographical references and index.
    ISBN 1- 56080-089-5 (vol.)
    1. Seismic prospecting. I. Galbraith, Mike. II. Peirce, John. III. Hardage, Bob Adrian, 1939- IV. Title. V. Geophysical development series ; v. 9.

TN269.8 .C67 2000
622'.1592--dc21

                                                                     00-026159

ISBN   0-931830-41-9 (Series)
ISBN   1-56080-089-5 (Volume)

Society of Exploration Geophysicists
P.O. Box 702740
Tulsa, OK 74170-2740

©2000 Society of Exploration Geophysicists
All rights reserved. This book or parts hereof may not be reproduced in any form without written permission from the publisher.

Published 2000
Reprinted 2008
Reprinted 2011

Printed in the United States of America

Acknowledgments

The authors wish to acknowledge the contributions from the staff members of Geophysical Exploration & Development Corporation and Seismic Image Software Ltd., both of Calgary, Alberta, Canada.

Several other individuals made significant contributions to this volume: in particular, Peter Eick of Conoco, who added a lot of practical perspective (and photographs), Gijs J. O. Vermeer of 3DSymSam, and Malcolm Lansley of Western Geophysical, who reviewed the text at an earlier stage and had many suggestions, which, in part, have been incorporated.

We also thank the people who have taken our 3-D design courses and who have contributed to this book through their numerous discussions with us. With their help, many aspects of 3-D design became more clearly understood.

Introduction

This book is intended to give readers the tools to start designing 3-D seismic surveys. The substantial experiences of the authors in designing and acquiring land 3-D seismic surveys make this a practical book.

Readers are expected to have a general working knowledge of 2-D seismic data acquisition, processing, and interpretation. Some 3-D experience is helpful but not necessary to understand this material. Practical exercises are included to facilitate understanding of the subject matter.

Throughout this book, an attempt has been made to provide data examples in both metric and imperial units. The imperial examples are shown in *italics* and are not necessarily a straight conversion from the metric examples, but rather are parameter values that are in common usage.

Some geophysicists may want to enhance their knowledge of 3-D design by reading papers concerning their particular special interests. A reference list and a collection of other recommended reading are provided at the end of the book.

Abbreviations

The commonly used abbreviations listed here will be used in this text. Other abbreviations may be used throughout the text and are explained when used.

AVA	amplitude variation with angle of incidence
AVO	amplitude variation with offset
B	bin dimension
B_r	in-line bin dimension (receiver direction)
B_s	cross-line bin dimension (source direction)
CCP	Common Conversion Point
CMP	Common Mid Point
DMO	Dip MoveOut
f	frequency
f_{dom}	dominant frequency
f_{max}	maximum frequency
FT	fold taper
FZ	Fresnel zone
k	velocity factor (increase of velocity with depth)
K	wavenumber
MA	migration apron
NC	number of channels
NMO	Normal MoveOut
NRL	number of receiver lines
NSL	number of source lines
p	ray parameter
R_F	radius of first Fresnel zone
RI	receiver interval
RLI	receiver line interval
RLL	receiver line length within the patch (i.e. in-line patch dimension)
SD	source density or number of source points per unit area
SI	source interval
SLI	source line interval
t	two-way traveltime
T	period; temporal wavelength
TMA	total migration apron
U	unit factor
V_{ave}	average velocity from surface to the reflecting horizon
V_{int}	interval velocity immediately above the reflecting horizon
V_0	velocity at surface
V_z	velocity at depth Z
X_{max}	maximum recorded offset
X_{min}	largest minimum offset
X_{mute}	depth varying mute distance
X_r	in-line dimension of the patch
X_s	cross-line dimension of the patch
Z	depth to reflecting horizon
λ	wavelength
λ_{dom}	dominant wavelength
λ_{max}	maximum wavelength
θ	geologic dip angle
θ_o	take-off angle

Conversion Tables

To convert from imperial units	to metric units	multiply by
Inches (in)	Centimeters (cm)	2.54
Feet (ft)	Meters (m)	0.3048
Miles (mi)	Kilometers (km)	1.609
Square miles (mi^2)	Square kilometers (km^2)	2.56
Acres (ac)	Hectares (ha)	0.405
Barrels (bbls)	Cubic meters (m^3)	0.159
Thousand cubic feet gas (mcf)	Thousand cubic meters (10^{3}m^3)	0.028169
Pounds (lb)	Kilograms (kg)	0.454
To convert from metric units	**to imperial units**	**multiply by**
Centimeters (cm)	Inches (in)	0.3937
Meters (m)	Feet (ft)	3.28
Kilometers (km)	Miles (mi)	0.6215
Square kilometers (km^2)	Square miles (mi^2)	0.39
Hectares (ha)	Acres (ac)	2.47
Cubic meters (m^3)	Barrels (bbls)	6.29
Thousand cubic meters (10^{3}m^3)	Thousand cubic feet gas (mcf)	35.5
Kilograms (kg)	Pounds (lb)	2.2
To convert	**to**	**multiply by**
Thousand cubic feet gas (mcf)	Barrel of oil equivalent (boe)	0.1
Miles (mi)	Feet (ft)	5280
Square miles (mi^2)	Acres (ac)	640
Square miles (mi^2)	Hectares (ha)	256
Hectares (ha)	Square meters (m^2)	10 000

Table of Contents

CHAPTER 1   INITIAL CONSIDERATIONS

1.1   Management Attitudes   1
1.2   Objectives   1
1.3   Industry Trends   2
1.4   Financial Issues   2
1.5   Target Horizons   5
1.6   Sequence of Events for Data Acquisition   5
1.7   Environment and Weather   8
1.8   Special Considerations of 3-D versus 2-D Data Acquisition   8
1.9   Definitions of 3-D Terms   8
Chapter 1 Quiz   12

CHAPTER 2   PLANNING AND DESIGN

2.1   Survey Design Decision Table   13
2.2   Orthogonal Geometry   14
2.3   Fold   14
2.4   In-line Fold   16
2.5   Cross-line Fold   17
2.6   Total Fold   18
2.7   Fold Taper   19
2.8   Signal-to-Noise Ratio (S/N)   20
2.9   Bin Size   20
    2.9.1   Target Size   21
    2.9.2   Maximum Unaliased Frequency   21
    2.9.3   Lateral Resolution   24
        2.9.3.1   Lateral Resolution after Migration   25
        2.9.3.2   Separation of Diffractions   26
    2.9.4   Vertical Resolution   27
Let's Design a 3-D—Part 1   28
2.10 X_{min} 29
2.11 X_{max} 32
    2.11.1   Target Depth   34
    2.11.2   Direct Wave Interference   34
    2.11.3   Refracted Wave Interference (First Breaks)   35
    2.11.4   Deep Horizon Critical Reflection Offset   35
    2.11.5   Maximum NMO Stretch to Be Allowed   35
    2.11.6   Required Offset to Measure Deepest LVL (refractor)   35
    2.11.7   Required NMO Discrimination   35
    2.11.8   Multiple Cancellation   35
    2.11.9   Offsets Necessary for AVO   35
    2.11.10   Dip Measurements   35

Let's Design a 3-D—Part 2   36
Chapter 2 Quiz   37

CHAPTER 3   PATCHES AND EDGE MANAGEMENT

3.1   Offset Distribution   39
3.2   Azimuth Distribution   40
3.3   Narrow versus Wide Azimuth Surveys   40
3.4   85% Rule   41
3.5   Fresnel Zone   46
3.6   Diffractions   47
3.7   Migration Apron   47
3.8   Edge Management   48
3.9   Ray-Trace Modeling   51
3.10 Record Length   51
Let's Design a 3-D – Part 3   53
Let's Design a 3-D – Summary   54
Chapter 3 Quiz   55

CHAPTER 4   FLOWCHARTS, EQUATIONS, AND SPREADSHEETS

4.1   3-D Design FlowChart   57
4.2   Basic 3-D Equations—Square Bins   57
4.3   Basic 3-D Equations—Rectangular Bins   59
4.4   Basic Steps in 3-D Layout—Five-Step Method   59
4.5   Graphical Approach   61
4.6   Standardized Spreadsheets   62
4.7   Estimating the Cost of a 3-D Survey   69
4.8   Cost Model   69

CHAPTER 5   FIELD LAYOUTS

5.1   Full-Fold 3-D   77
5.2   Sampling the 5-D Prestack Wavefield   77
5.3   Swath   81
5.4   Orthogonal   81
5.5   Brick   83
5.6   Nonorthogonal   83
5.7   Flexi-Bin® or Bin Fractionation   85
5.8   Button Patch   89
5.9   Zig-Zag   91
5.10 Mega-Bin   92
5.11 Hexagonal Binning   93
5.12 Star   96

5.13   Radial   96
5.14   Random   96
5.15   Circular Patch   99
5.16   Nominal Fold Comparison   99
Chapter 5 Quiz   105

CHAPTER 6   SOURCE EQUIPMENT

6.1   Explosive Sources   107
6.2   Dynamite Testing   113
6.3   Dynamite Shooting Strategy   113
6.4   Vibrators   113
6.5   Vibrator Array Concepts   114
6.6   Vibrator Testing   116
6.7   Vibrator Deployment Strategy   118
6.8   Other Sources   119
Chapter 6 Quiz   119

CHAPTER 7   RECORDING EQUIPMENT

7.1   Receivers   121
7.2   Recorders   123
7.3   Distributed Systems   124
7.4   Telemetry Systems   126
7.5   Remote Storage   127
Chapter 7 Quiz   127

CHAPTER 8   ARRAYS

8.1   The Question of Arrays   129
8.2   Geophone Arrays   129
8.3   Source Arrays   131
8.4   Combined Array Response   131
8.5   Stack Arrays   131
8.6   Hands-Off Acquisition Technique   134
8.7   Symmetric Sampling   134

CHAPTER 9   PRACTICAL FIELD CONSIDERATIONS

9.1   Surveying   135
9.2   Script Files   138
9.3   Templates   141
9.4   Roll-On/Off   141

x   Table of Contents

9.5   No Roll-On/Off   142
9.6   Swath Width   142
9.7   Shooting Strategy   143
    9.7.1   Vibrator   145
    9.7.2   Dynamite   146
9.8   Large Surveys   146
9.9   Field Visits (QC)   147
9.10   Offsets and Skids   148
9.11   General Considerations   148
    9.11.1   Imaging Area   148
    9.11.2   Cables   149
    9.11.3   Permitting   150
    9.11.4   Safety   150
9.12   Field Examples   151
9.13   Field QC (Data)   152
    9.13.1   Positional Data Quality   152
    9.13.2   Seismic Data Quality   152
    9.13.3   Verify Seismic/Positional Data Relationship   153
Chapter 9 Quiz   155

CHAPTER 10   PROCESSING

10.1   Processing   157
10.2   Processing Stream   157
10.3   Refraction Statics   158
10.4   Velocity Analysis   158
10.5   Reflection Statics   161
10.6   Dip Moveout   162
10.7   Stack   166
10.8   Acquisition Footprints   166
10.9   Migration and Random Sampling   167
10.10   Making Adjustments for Data Quality   168
Chapter 10 Quiz   169

CHAPTER 11   INTERPRETATION

11.1   Interpretation Systems   171
11.2   Mapping   171
11.3   Integration   172
11.4   Acquisition Footprints   173
11.5   Seismic Attributes   173
11.6   Geostatistics   173
11.7   Immersion Technology   173

CHAPTER 12 SPECIAL INTEREST TOPICS

12.1 Digital Orthomaps and GIS data   175
12.2 Transition Zones   175
12.3 Prestack Time and Depth Migration   176
12.4 Time-Lapse (4-D) Seismic   177
12.5 Converted Wave 3-D Design   177
12.6 3-D Inversions   180
12.7 Future Directions   181

ANSWERS TO QUIZ QUESTIONS   183

GLOSSARY   185

REFERENCES   195

INDEX   201

Initial Considerations 1

1.1 MANAGEMENT ATTITUDES

The management of an oil company needs to be familiar with the acquisition, processing, and interpretation requirements that a 3-D survey may place on its staff. If a company's management has had prior exposure to 3-D seismic surveys, less education by the technical staff (usually the geophysicists) is necessary before recommending a 3-D survey. There may be some preconceived ideas as to the final products that might be delivered at various stages. It is important to emphasize that success or failure in a past 3-D survey may not necessarily be duplicated in future programs. Modifying the design, acquisition, and processing parameters can make significant improvements. Conversely, results may be less than expected if poor design parameters are chosen.

Geophysicists may find themselves serving one or more customers. Once 3-D data have been acquired and interpreted, the interpreted data set will become a focal point for several people because the interpretation will be delivered to team members that practice different disciplines (Figure 1.1). The data also become a valuable asset with resale value.

Possible partners may need to be informed at an early stage about the planned operations so they can set aside the anticipated financial and personnel resources. They may wish to have significant input into choosing the area for the 3-D survey, or in planning the design, or they may wish to contribute in some other manner. Obtaining their approval is much easier if they have been intimately involved from the start. This approach gives partners a sense of ownership. Sometimes the company that operates the field is not the one that contributes most to a 3-D survey. It is possible, for example, that another partner in the area could operate an extensive seismic program. Information exchange is an important as-

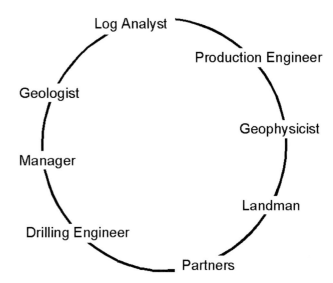

Fig. 1.1. The geophysicist as part of the exploration/exploitation team.

pect of doing the very best technical job in 3-D design and acquisition.

1.2 OBJECTIVES

A company needs to establish early and clearly why a 3-D survey is to be recorded (some possible reasons are listed in Figure 1.2). These goals must be kept in mind during all phases of the planning process. Any seismic program must be planned, recorded, processed, and interpreted in time to deliver sufficient results to the owners of the data so that they can evaluate all results along with other information and constraints that they may have.

Most of the reasons for recording the 3-D seismic data listed in Figure 1.2 do not need any explanations.

2   Initial Considerations

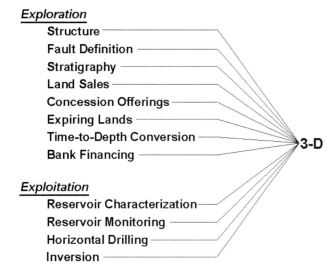

Fig. 1.2. Different reasons for shooting a 3-D seismic survey.

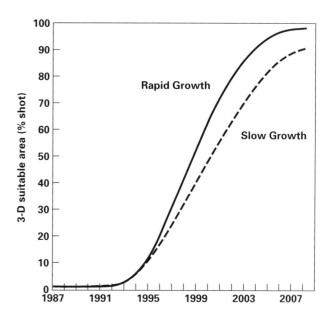

Fig. 1.3. Application of 3-D seismic technology to areas suitable for 3-D data acquisition in North America (Koen, 1995 after A. Cranberg, Aspect Management Corp.).

For example, reservoir monitoring may be essential for better production practices in large fields. The differences observed in 3-D seismic surveys recorded over the same field with a separation period of several years show the progress of depletion and flooding practices. Such "4-D" or "time-lapse" surveys are becoming more common.

1.3   INDUSTRY TRENDS

Both large and small companies now use high-technology tools to obtain data improvements. This practice is particularly true for the energy sector, especially regarding the use of 3-D seismic technology. Success ratios for oil companies have been increased by using 3-D data. In a worldwide study, one large company registered an increase in their success rate from 13% in 1991 using 2-D data to 44% in 1996 using 3-D technology extensively (Ayler, 1997). While the success rate using 2-D data alone has remained constant, the success rate using 3-D seismic technology has shown a dramatic improvement.

Small independent companies may acquire small 3-D surveys to help detail relatively small land holdings surrounding existing production. Larger oil companies may acquire 3-D data over larger areas of 10s or 100s of km^2. Often these surveys are for exploratory purposes only. One important new trend is that acquisition contractors are now offering to record huge 3-D participation surveys not only in the offshore environment where the practice has been done for a while, but also onshore.

One estimate for North America indicates that by the year 2007, essentially all of the area in the US and Canada that is suitable for the application of 3-D seismic technology will be covered (Figure 1.3). Since this is an average estimate, there will be many areas that will be covered more than once by a 3-D survey without the intention of doing 4-D surveys. With acquisition prices falling at a fast rate and higher channel capacities being available, 3-D acquisition becomes the favored choice over 2-D acquisition.

Many major oil companies have the necessary resources and expertise to plan, acquire, process, and interpret 3-D data in-house, while medium and small-size oil companies rely on the knowledge and experience that consultants offer. By constantly dealing with the subject of 3-D data acquisition, it is much easier to be proficient at planning and operating such surveys.

1.4   FINANCIAL ISSUES

Cost factors play an important role in making decisions about the expenditures for a 3-D survey. The exploration team must prove to management that a dense grid of geophysical data tied to geological information from existing wells provides significant economic benefits by reducing the number of dry holes and overall costs. In the past, at least one discovery well in a particular prospect area was needed to convince management to spend additional resources on 3-D seismic data. Recently there has been a trend to use 3-D technology even in a purely exploratory environment. The cost of acquiring several 2-D programs possibly spread over many years may be just as high as a 3-D survey. In addition, the problems of interpreting and consistently incorporating various vintages of 2-D data lead to inevitable uncertainties that may be insurmountable. Therefore, acquiring 3-D seismic data provides a more cost-effective evaluation of a prospect by providing increased confidence in the seismic interpretation and new technical information.

Budget constraints need to be made clear at the early planning phase; otherwise, unrealistic designs may result. If the budget numbers are too low, the 3-D survey may be under-designed and unable to meet management expectations. On the other hand, if the budget numbers are too high, the designers of the 3-D survey may over-design in areal extent or in other technical specifications. Important considerations are: who ultimately controls the budget; who approves any unanticipated changes, especially cost overruns; does the planning committee meet on particular dates or at irregular intervals; and how difficult is it to obtain timely approvals in order to maintain the time schedule?

Management needs to be able to evaluate the economic rate of return for any project in question. The potential of a prospect and its associated risk must warrant the cost of a 3-D seismic survey. More often than not, a 3-D survey is difficult to justify on a single-well basis. However, a 3-D seismic survey may very well be worthwhile if dry holes are to be avoided. Exploration wildcat wells are commonly that type of situation. A 3-D survey may make a wildcat less "wild" and result in the drilling of significant discoveries. The cost of missed opportunities is very high.

On a project that has numerous development locations, and even for low-cost drilling of relatively shallow wells, 3-D seismic surveys are often economically justifiable. If many step-out wells and in-fill locations are anticipated, project economics may dictate a 3-D survey. Similarly, plans for horizontal drilling may require tightly controlled seismic data. For example, if the target horizon is relatively thin, drilling engineers may need high-resolution sampling in all three dimensions to keep the drill bit in the reservoir.

Even a small increase in the success ratio of drilling wells with a 3-D survey (e.g., 1:5) versus without a 3-D survey (e.g., 1:6) could justify the cost of a 3-D survey (Figure 1.4). Assume drilling 6 wells at a dry hole

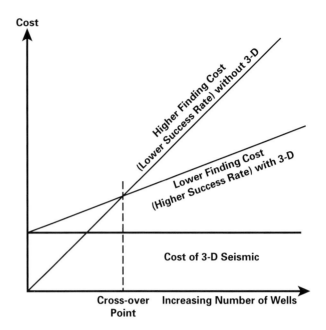

Fig. 1.4. Project economics for wells and a 3-D seismic survey.

cost of $500,000 each versus drilling 5 wells at a cost of $500,000 each with 3-D data.

$$6 \times \$500K = \$3,000,000 \quad \text{without 3-D}$$
$$5 \times \$500K = \$2,500,000 \quad \text{with 3-D}$$
$$\text{difference} = \$\ 500,000$$

This example indicates that one might have $500,000 available to invest in the acquisition of a 3-D data set. Under current economic conditions, this amount of money might pay for 10 mi^2 or more of 3-D seismic data.

Head (1998) provides a more thorough analysis of the value of 3-D data through decision tree analysis that can guide the explorationist in his or her decisions (Figure 1.5). There are numerous decision points when deciding whether to drill and/or whether to acquire additional 2-D or 3-D seismic data. One can assign certain probabilities to particular exploration results that can be achieved with 3-D seismic data by using expected value concepts (Figure 1.6, Table 1.1). Using the terminology in Table 1.1, the probability of an economic success P_{es} is

$$P_{es} = P_{source} \times P_{migration} \times P_{reservoir} \times P_{trap}, \quad (1.1)$$

and the expected monetary value EMV is

$$\text{EMV} = \text{NPV}_{success} \times P_{es} + \text{NPV}_{failure} \times P_{ef}. \quad (1.2)$$

4   Initial Considerations

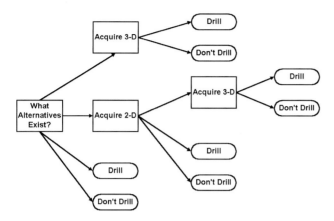

Fig. 1.5. Decision tree analysis to guide the exploration decision process.

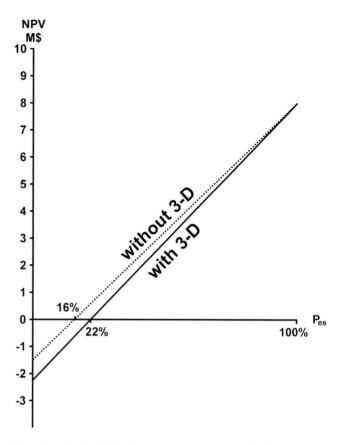

Fig. 1.6. Probability of economic success (P_es) versus net present value (NPV).

Through such analysis one can establish the maximum economic value to ascribe to a 3-D seismic survey. The difference in the expected monetary value (EMV) of the project without and with 3-D data determines the maximum amount that can be spent on a 3-D survey. Total project economics, cost of money, and a possible increase or decrease in the total project NPV are not taken into account in Figure 1.6. Aylor (1995) points out that many 3-D surveys add value to exploration and development projects because more wells can be drilled.

Costs for 3-D surveys vary depending on the area where the survey is to be conducted, the availability of equipment and crews, and the complexity of the geography. In general, one can expect to pay on the order of $10 000 to $50 000 per km^2 for data acquisition. High-resolution work for smaller bins and high-fold surveys can exceed those costs. On the other hand, sparse 3-D surveys (Bouska, 1995) or very focused 3-Ds (Servodio et al., 1997) can provide cost savings. An analysis of the economics of different acquisition parameters is essential in evaluating whether high S/N ratios and small bin sizes are warranted. Such analysis is possible by decimating exist-

Table 1.1 Profitability Table

		w/o 3-D	with 3-D	% change
P_{source}	Probability of hydrocarbon source	90%	90%	0%
$P_{migration}$	Probability of migration of hydrocarbons	80%	80%	0%
$P_{reservoir}$	Probability of reservoir/porosity	70%	80%	14%
P_{trap}	Probability of seal/trap	30%	40%	33%
$NPV_{success}$	Net present value of successful well	$8,000,000	$8,000,000	
$NPV_{failure}$	Net present value of dry well	($1,500,000)	($1,500,000)	
P_{es}	Probability of economic success	15%	23%	52%
P_{ef}	Probability of economic failure	85%	77%	9%
EMV	Expected monetary value	($63,600)	$688,800	
VOI	Value of information (e.g., 2-D, 3-D, interpretations)		($752,400)	
	Required success ratio for single well	16%	22%	39%

ing data sets and interpreting the individual data sets separately (Schroeder and Farrington, 1998).

Processing costs vary but are usually in the range of 5–10% of acquisition. A detailed interpretation should be in the same cost range as processing.

1.5 TARGET HORIZONS

A 3-D seismic survey should be designed for the main zone of interest (primary target). This zone will determine project economics by affecting parameter selection for the 3-D seismic survey. Fold, bin size, and offset range all need to be related to the main target. The direction of major geological features, such as faults or channels, may influence the direction of the receiver and source lines.

Secondary zones or other regional objectives may have a significant impact on the 3-D design as well. A shallow secondary target, for example, may require very short near offsets. Deeper regional objectives and migration considerations may dictate that the far offset of the survey be substantially greater than the maximum stacking offset used in the fold calculation at the target level (Figure 1.7).

1.6 SEQUENCE OF EVENTS FOR DATA ACQUISITION

Preparing an overall time line for data acquisition will avoid surprises and keep expectations somewhat close to reality. This time planning should also help in meeting critical deadlines such as land sales, lease expirations, or bid submission dates. The technical team should update this time line as the project progresses, so that the parties involved are kept abreast of the changes. A realistic time line needs to be established early so that expectations are on track with the overall process of obtaining the data (Figure 1.8). The time required for each step in the time line varies widely from area to area. A small 3-D survey can be completed from scouting to drilling within 6 to 8 weeks, while larger surveys in difficult access areas may demand two years or more. In-depth knowledge of local time requirements is essential.

A scouting trip to the 3-D area may provide substantial information for the design of the 3-D survey; e.g., existing cut lines may dictate line intervals and/or direction, or surface cover could influence dynamite hole depth and charge size. All technical parameters must be kept in mind when designing a

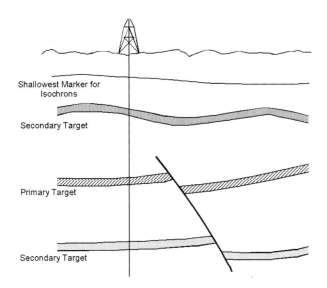

Fig. 1.7. Primary target horizon versus secondary targets.

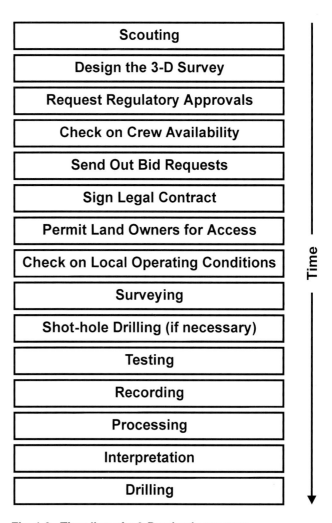

Fig. 1.8. Time line of a 3-D seismic program.

survey. The design may need to be updated as more elements and parameters of the time line become known. Operators should request all necessary regulatory approvals and stay in close contact with regulatory bodies to ensure smooth operation, remembering to consider past survey requirements and costs such as forestry regulations, damages, reseeding, and correction of erosion problems.

Critical questions such as "Are experienced 3-D crews available locally?" need to be answered early in the project. If crews need to be shipped across the country or even from another country or continent, major delays should be expected, especially when clearing customs. Some 3-D seismic equipment is often hard to get through customs because officials often do not understand the technology. Knowing the availability of spare parts is important if anything goes wrong in the field. For example, if cables become damaged, how much replacement equipment can be brought in and how long will it take?

Other key questions that need to be considered are: What kinds of data acquisition bid procedures are customary at the prospect location? How much time is involved from requesting such bids to their actual submission? How much time do the contractors need to put a reasonably accurate bid together? Do contractors need to research local conditions, and to what extent? The oil company may have bidding procedures laid out in a very particular manner, e.g., bids may need to be presented in a form of cost/km^2, cost/source point, cost/day, or total project cost, just to name a few variations. The content requirements for an acquisition bid should be clearly known to everyone involved in the bidding process. If the contractor has to sign a standard contract before embarking on a job, the oil company may want to include this contract at the stage of requesting a bid. It is important to negotiate a satisfactory contract that meets the needs of the oil company's particular situation and that also reflects the political environment.

Many acquisition contractors will subcontract parts of the job such as surveying, shot-hole drilling, and bulldozer work. The costs of these subcontracts are usually considered extras and therefore may not be included in the overall cost/unit basis. A best effort must be made to estimate the extent of these extras to arrive at a realistic total cost figure for a 3-D seismic survey. These so-called extras may more than double the acquisition cost. The uncertainties in cost because of allowances for bad weather can also represent a significant portion of the total cost.

Turnkey quotes help set the price for most of the acquisition costs. This bidding policy assures a client that the crew will work fast. Some element of supervision must be introduced to obtain the required quality of service. Daily rates, on the other hand, do not give the crew an incentive to work fast. However, if no other jobs are waiting for the crew, the best level of service possibly can be obtained via a daily rate option.

Some companies may choose to hire an acquisition crew for an entire acquisition season or even for several years. Under such circumstances, the need to negotiate every seismic program disappears and better planning needs to be implemented for continuous operation. The price guarantee that usually accompanies such arrangements is a big advantage over the uncertainties that industry experiences in fluctuating markets.

Often the legal contract that a contractor provides is not comprehensive. If any field problems, accidents, or insufficiencies arise, an incomplete contract may limit the legal protection for the acquisition contractor and the client. It is advisable to have legal representatives review the contract and ensure that sufficient protection exists for both parties. If experience with such contracts does not exist within the organization, then outside advice should be requested.

Permits may be required from land surface owners to obtain entry. Such permits should be requested as early as possible because permit issues can affect a 3-D survey in a number of different ways. Permitting may need to be started significantly earlier in the time line of Figure 1.8. Landowners may not want to see any member of an acquisition crew during the growing season, even if crop damages are to be paid. Slight changes in the design or layout of the receivers and sources may make a huge difference to particular landowners. For example, by moving a portion of a line across the fence to neighboring noncrop land, one may avoid crop damages and pay another landowner permit fees. This is a beneficial scenario for all concerned. Good rapport with landowners will go a long way to assuring access to their lands, and keeping damages to a minimum will help the next seismic crew that wants to work in the same area. Often permitting by km^2 is less expensive than permitting by line km. Permitting by area also gives more freedom of choice in the field.

If a landowner controls a high percentage of the lands within a 3-D survey and is opposed to the seismic operation, the entire program may be in jeopardy. Large gaps of "no coverage" on a 3-D survey are undesirable, and such opposition may cripple the planned survey and perhaps cancel part of the exploration program.

In at least one U.S. state (e.g., Texas), it is illegal to record a geophysical measurement of any kind over another owner's mineral rights without a permit from that owner (geophysical trespass). There is considerable confusion as to how the relevant laws should be interpreted. Most operators are now being diligent to obtain permits over all relevant lands to protect themselves against possible liability. Many operators of 3-D surveys are trimming their surveys to ensure that there are no stacked traces over areas not covered by permits (e.g., not undershooting corners). Interested readers should see AAPG Explorer, June 1995, for a discussion of the Burr Ranch case and related issues.

Key questions to consider include the following: How much is known about the operating conditions? Which contractors have experience in the area to contribute to the successful operation of such programs? Will the contractor share this information at the time of bid submission or only if they win the contract?

If the knowledge of local operating conditions is limited or data quality to be expected is unknown, then a 2-D test line may be required and is sometimes essential for correct parameter selection. Source and receiver array tests are much easier to conduct, especially for small 3-D surveys, when a 2-D test line is being considered as a first evaluation of the prospect. On a large 3-D survey, it might be justified to conduct the required tests at the start of the survey. Large surveys may require a variety of sources or receivers (such as in a transition zone), and several test sequences may be performed throughout the survey. Sometimes the local conditions are known well enough that a 3-D survey may proceed without any testing at all.

Surveyors need to go to the field and establish a perimeter of the 3-D seismic survey before filling in the specific source and receiver lines. Global Positioning Systems (GPS) technology provides good survey accuracy and is a faster surveying technique than more traditional Electronic Distance Measurement (EDM) devices. Differential GPS, which relies on a well-established local base station, offers even greater accuracy; its horizontal accuracy is <1 m while the vertical accuracy is 2 to 3 m presently. Accuracies of a few centimeters can be achieved with longer time on each station. Once the grid has been established, chainers will mark every source and receiver group location. GPS does not work well in dense tree cover or in deep ravines where satellites cannot be seen. For further GPS information, see Harris and Longaker (1994).

Shot-hole drilling may commence immediately following, or even concurrent with, the surveying. Usually the entire 3-D area will be drilled before the recording crew arrives, assuming that all source parameters have been previously established. This drilling schedule reduces noise interference between the drilling units and the data recording. There is also no chance of the drillers being in the way of the layout crew.

Vibrator trucks or buggies may start sweep production once the sweep parameters have been tested. It is always advisable to complete phase, peak force, and correlation tests before going into production mode. These tests should be repeated several times during the acquisition program and should be done more than once daily.

If a program does not allow or warrant any testing, an attempt should be made to find past tests or data. Testing is important for both dynamite and vibrator (or any other source) data acquisition. Such tests may be instructive for any future programs, even if the present survey cannot benefit from the test results.

The recording crew places receivers on the ground in a predetermined array. The geophones are connected in receiver groups, which then transmit the digital information in a variety of ways to the recorder. Cable-based distributed recording systems require a continuous cable connection from the geophones to the recorder, thus one can walk out the cable connections from any geophone all the way back to the recording truck. An alternate technique for data transmission is the telemetry system, which uses radio signals to transmit data to the recording truck instead of cables. In the case of the I/O RSR system, data are recorded locally and then retrieved periodically for storage on tape. For this system, the radio is a control unit for initiation of recording and quality control, but it is not used for data transmission.

The recorder unit (dog house) has a complete set of electronics that allows data correlation (for vibrators) and the recording and display of shot records that show traces corresponding to all geophone groups.

Some crews operate on a 24-hour basis to reduce the overhead cost per source point. One has to check whether local customs and/or laws and safety concerns allow such around-the-clock operation. Limiting data-recording activity to dawn-to-dusk significantly lengthens the number of days needed to record a 3-D seismic survey.

Field tapes are sent to the processor for analysis and imaging of the data, or more recently, the data can be processed in the field. The choice of the processor should be decided before the crew enters the field. Survey notes need to be reduced to final coordinates, and the final survey geometry must be forwarded to the processor.

Interpretation on paper and/or on a workstation usually gives a clear idea of the geological variations in the area of the 3-D survey. Drilling of exploration or development wells should commence only after sufficient time has been allotted for a thorough and complete interpretation.

1.7 ENVIRONMENT AND WEATHER

Environmental issues play an important role in today's world—especially in seismic data acquisition. One should protect the environment as much as possible during all field operations. Line cutting in forested areas should be limited to the smallest width necessary. Small jogs in the lines are often requested to protect the pristine appearance of the woods and to protect wild life. In mountainous areas or other difficult terrain, helicopter support may be essential for shot-hole drilling or for laying receiver cables to minimize damaging effects on the environment.

Wildlife protection issues have to be addressed such as mating seasons and migration paths. In a transition zone, fish spawning might be a concern at certain times of the year. In some parts of the world, rodents may chew cables and hinder successful data transmission. The use of wooden pegs for station flags lessens the damage to farm equipment and animals, such as cows, which often chew station flags and cables. Some areas are so environmentally sensitive that local interest groups may lobby government officials to prevent any seismic operation, or they may interfere directly with seismic or drilling operations. Recent trends in the environmental industry have shown a recognition that stopping the seismic operations also effectively stops oil and gas exploration; therefore, good community relations in advance of operations are a wise investment in time and effort.

Weather conditions may constrain operation of a seismic program to certain times of the year. Rain or snow may alter ground conditions to such a degree that data quality is severely diminished. Crew movement may also be hindered. In cold climates, one may want to wait for frost before laying out geophones to improve the coupling of the geophone spikes to the ground and to minimize surface damages. Snow cover may need to be removed to allow the frost to enter the ground rather than allowing the snow to act as an insulator. Often receiver lines need to be cleared several times if significant new snowfall occurs during crew operations. In warmer climates, extreme heat conditions may hinder the effectiveness of the crew personnel and may pose a serious safety hazard.

1.8 SPECIAL CONSIDERATIONS OF 3-D VERSUS 2-D DATA ACQUISITION

One needs to specify the objectives of a 3-D survey more precisely than for a 2-D survey because the acquisition parameters are more difficult to change in mid-program. For example, with a 3-D survey (as opposed to 2-D) much more line cutting is required in forested areas. This makes it harder to obtain approval from regulatory bodies, and even when approval is granted, one may be limited to using existing cut lines or be restricted to hand-cut receiver lines, which can slow the operation. On a 3-D survey, the equipment stays on the ground much longer than in 2-D data programs. This factor exposes 3-D equipment to more environmental, vehicular, weather, theft, and wildlife damage.

Spatial sampling in 3-D programs is usually much coarser than in 2-D programs (e.g., 20 to 40 m bins in a 3-D survey versus 5 to 15 m trace spacings in 2-D surveys). It is important to decide whether this coarser sampling is sufficient to resolve structural dips and to properly image geological features. For 2-D lines, linear source and geophone arrays are the norm. The effects that source-receiver azimuths have on these geophone (or source) arrays is a topic of recent papers and research. There is no consensus yet in the industry on the type of arrays to use in 3-D data acquisition.

Finally, 3-D sources and receivers are laid out over an area, and 3-D recordings have an azimuthal element that is not present in 2-D efforts. Good azimuthal distributions are usually, but not always, desirable. If any out-of-the-plane phenomenon exists in a 2-D profile, often one is unable to determine the direction of its cause. In contrast, 3-D migration has a better chance of properly positioning such anomalies.

One can argue at length about various aspects of 3-D versus 2-D imaging. 3-D data have a common set of acquisition and processing parameters over a large area and therefore are easier to interpret than a series of 2-D lines of various vintages. A 3-D data volume is continuous, and one may extract profiles in any direction out of this volume. However, in some situations, 2-D data may be more beneficial than 3-D data, e.g., when there is a need to gain a regional perspective or to improve local resolution with a small trace spacing.

1.9 DEFINITIONS OF 3-D TERMS

Figures 1.9 and 1.10 show a plan view of an orthogonal 3-D survey that illustrates most of the terminology used in this book.

1.9 Definition of 3-D Terms   9

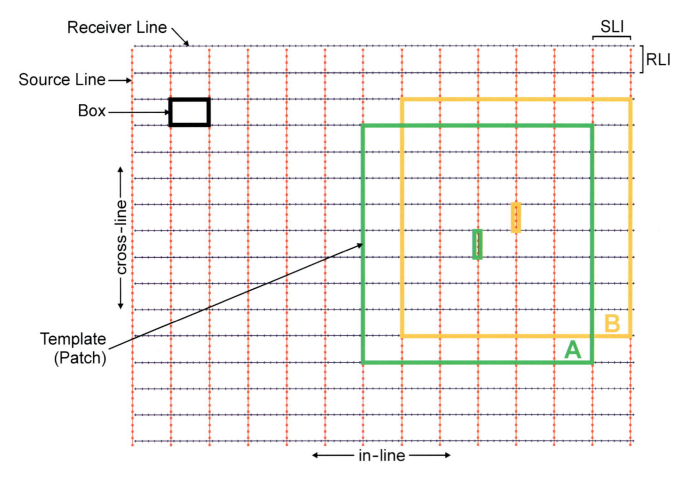

Fig. 1.9. 3-D survey layout terms.

Note: This book uses SI notation as a standard; however, most numerical examples are presented in imperial units as well, and these are printed in *italics*. Rather than directly converting metric units to imperial units, we have chosen the natural imperial units that would be used in the particular situation (e.g., a 30 m bin size might be equivalent to a *110 ft* bin size).

Box (sometimes called "Unit Cell") In orthogonal 3-D surveys, this term applies to the area bounded by two adjacent source lines and two adjacent receiver lines (Figures 1.9 and 1.12). The box usually represents the smallest area of a 3-D survey that contains the entire survey statistics (within the full-fold area). In an orthogonal survey, the midpoint bin located at the exact center of the box has contributions from many source-receiver pairs; the shortest offset trace belonging to that bin has the largest minimum offset of the entire survey. In other words, of all the minimum offsets in all CMP bins, the minimum offset in the bin at the center of the box has the biggest X_{min}. Different layout strategies attempt to deal with this concept in a variety of ways.

CMP Bin (or Bin) A small rectangular area that usually has the dimensions $(SI \div 2) \times (RI \div 2)$. All mid-

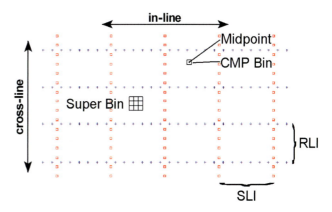

Fig. 1.10. 3-D survey bin terms.

points that lie inside this area, or bin, are assumed to belong to the same common midpoint (Figure 1.10). In other words, all traces that lie in the same bin will be CMP stacked and contribute to the fold of that bin. On occasion, one may choose the area over which traces are stacked to be different from the bin size in order to increase stacking fold. This introduces some data smoothing and should be performed with caution because it affects spatial resolution.

Cross-line Direction The direction that is orthogonal to receiver lines.

Fold The number of midpoints that are stacked within a CMP bin. Although one usually gives one average fold number for any survey, the fold varies from bin to bin and for different offsets.

Fold Taper The width of the additional fringe area that needs to be added to the 3-D surface area to build up full fold (Figure 1.11). Often there is some overlap between the fold taper and the migration apron because one can tolerate reduced fold on the outer edges of the migration apron.

In-line Direction The direction that is parallel to receiver lines.

Midpoint The point located exactly halfway between a source and a receiver location. If a 480-channel receiver patch is laid out, each source point will create 480 midpoints. Midpoints will often be scattered and may not necessarily form a regular grid.

Migration Apron The width of the fringe area that needs to be added to the 3-D survey to allow proper migration of any dipping event (Figure 1.11). This width does not need to be the same on all sides of the survey. Although this parameter is a distance rather than an angle, it has been commonly referred to as the migration aperture. The quality of images achieved by 3-D migration is the single most important advantage of 3-D versus 2-D imaging.

Patch A patch refers to all live receiver stations that record data from a given source point in the 3-D survey. The patch usually forms a rectangle of several parallel receiver lines. The patch moves around the survey and occupies different template positions as the survey moves to different source stations.

Receiver Line A line (perhaps a road or a cut-line through bush) along which receivers are laid out at regular intervals. The in-line separation of receiver stations (receiver interval, RI) is usually equal to twice the in-line dimension of the CMP bin. Normally the field recorder cables are laid along these lines and geophones are attached as necessary. The distance between successive receiver lines is commonly referred to as the receiver line interval (or RLI). The method of laying out source and receiver lines can vary, but the geometry must obey simple guidelines.

Scattering Angle Assuming the presence of a point scatterer (diffraction point) at depth, the scattering angle is the angle between the vertical downgoing source-scatterer raypath and the upgoing scatterer-receiver raypath.

Signal-to-Noise Ratio The ratio of the energy of the signal over the energy of the noise. Usually abbreviated as S/N.

Source Line A line (perhaps a road) along which source points (e.g., dynamite or vibrator points) are taken at regular intervals. The in-line separation of sources (source interval, SI) is usually equal to twice the common midpoint (CMP) bin dimension in the cross-line direction. This geometry ensures that the midpoints associated with each source point will fall exactly one midpoint away from those associated with the previous source point on the line. The distance between successive source lines is usually called the source line interval (or SLI). SLI and SI determine the source point density (or SD, source points per square kilometer).

Source Point Density (sometimes called shot density), SD The number of source points/km^2 or *source points/mi^2*. Together with the number of channels, NC, and the size of the CMP bin, SD determines the fold.

Super Bin This term (and others like macro bin or maxi bin) applies to a group of neighboring CMP bins. Grouping of bins is sometimes used for velocity determination, residual static solutions, multiple attenuation, and some noise attenuation algorithms.

Swath The term swath, has been used with different meanings in the industry. First, and most commonly, a swath equals the width of the area over which source stations are recorded without any cross-line rolls. Second, the term describes a parallel acquisition geometry, rather than an orthogonal geometry, in which there are some stacked lines that have no surface lines associated with them.

Template A particular receiver patch into which a number of source points are recorded. These source

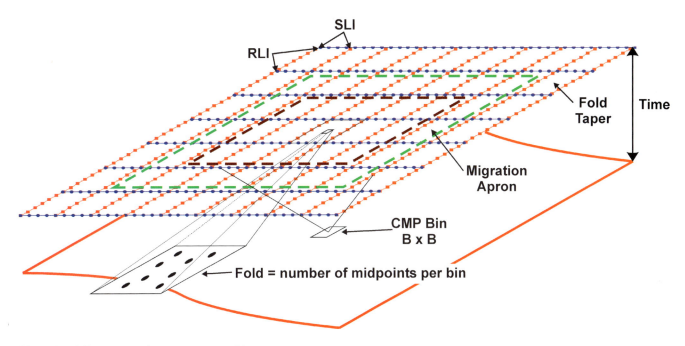

Fig. 1.11. 3-D survey edge management terms.

points may be inside or outside the patch. In equation form,

Template = Patch + associated source points.

X_{max} The maximum recorded offset, which depends on shooting strategy and patch size. X_{max} is usually the half-diagonal distance of the patch. Patches with external source points have a different geometry. A large X_{max} is necessary to record deeper events.

X_{min} The largest minimum offset in a survey (sometimes referred to as LMOS, largest minimum offset) as described under "Box." See Figure 1.12. A small X_{min} is necessary to record shallow events.

Assuming RLI and SLI of 360 m *(1320 ft)*, RI and SI of 60 m *(220 ft)*, the bin dimensions are 30 m × 30 m *(110 ft × 110 ft)*. The box (being formed by two parallel receiver lines and the orthogonal source lines) has a diagonal of:

$$X_{min} = (360^2 + 360^2)^{1/2} \text{ m}$$
$$= 509 \text{ m}$$

$$X_{min} = (1320^2 + 1320^2)^{1/2} \text{ ft}$$
$$= 1867 \text{ ft}$$

The value of X_{min} defines the largest minimum offset to be recorded in the bin that is in the center of the box. In this example the source and receiver stations are intentionally coincident at the line intersections for simplicity.

X_{mute} The mute distance for a particular reflector. Any traces beyond this distance do not contribute to

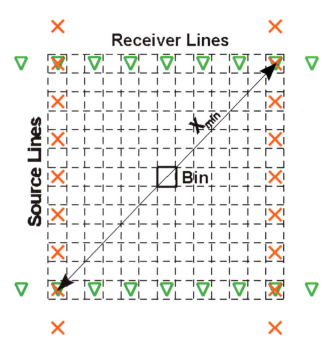

Fig. 1.12. X_{min} definition.

the stack at the reflector depth. X_{mute} varies with two-way traveltime.

Chapter 1 Quiz

1. Define receiver line interval.
2. What is migration apron?
3. How does one determine X_{min} in an orthogonal survey (orthogonal source and receiver lines)?
4. How large is a super bin?

2
Planning and Design

Survey design depends on so many different input parameters and constraints that it has become quite an art. Laying out lines of sources and receivers must be done with an eye toward the expected results. A solid understanding of the required geophysical parameters must be in place prior to embarking on a 3-D design project (Kerekes, 1998). Some rules of thumb and guidelines are essential to help one through the maze of different parameters that need to be considered. Computer programs are now available to assist in this task.

2.1 SURVEY DESIGN DECISION TABLE

Table 2.1 shares how to determine the values of fold, bin size, X_{min}, X_{max}, migration apron, fold taper, and record length that should be incorporated into a

Table 2.1 Survey Design Decision Table.*

Parameter	Definitions and Requirements
Fold	*Should be 3-D fold = 50% of 2-D fold (if S/N ratio is good) to 100% of 2-D fold (if higher frequencies are expected).* In-line fold = number of receivers × RI ÷ (2 × SLI). Cross-line fold = NRL ÷ 2.
Bin size	Use 3 to 4 traces across target. *Should be < V_{int} ÷ (4 × f_{max} × sin θ); for aliasing frequency.* Should provide N (= 2 to 4) points per wavelength of dominant frequency. *Lateral Resolution available: λ ÷ N or V_{int} ÷ (N × f_{dom}).*
X_{min}	*Should be less than 1.0 to 1.2 times depth of shallowest horizon to be mapped.*
X_{max}	*Should be approximately the same as target depth.* Should not be large enough to cause *direct wave interference,* refracted wave interference (first breaks), or *deep horizon critical reflection offset,* particularly in the cross-line direction, or intolerable *NMO stretch.* Should exceed offset required to see deepest LVL (refractor), offset required to cause NMO δt> one wavelength of f_{dom}, offset required to get *multiple discrimination* >3 wavelengths, and offset necessary for AVO analysis. Should be large enough to measure X_{max} as a function of dip.
Migration apron (full-fold)	*Must exceed radius of first Fresnel zone, diffraction width (apex to tail) for an upward scattering angle of 30°, i.e., Z tan 30° = 0.58 Z, and dip lateral movement after migration, which is Z tan θ.* Can overlap with fold taper.
Fold taper	Is approximately patch dimension ÷ 4.
Record length	Must be sufficient to capture target horizons, migration apron, and diffraction tails.

*The more important requirements are in *italics*. Additional definitions can be found in the Glossary.

14 *Planning and Design*

design. These parameters constitute the key design factors that need to be determined for a 3-D design. Each key parameter is discussed in this chapter and in Chapter 3.

2.2 ORTHOGONAL GEOMETRY

Often source and receiver lines are laid out orthogonal to each other in onshore 3-D surveys. Such an arrangement is easy for survey and recording crews to follow and keeping track of station numbering is straightforward. Receiver lines could run east-west and source lines north-south, as shown in Figure 2.1, or vice versa. This method is easy to lay out in the field and allows convenient equipment deployment ahead of shooting and roll-along operation. In this geometry, all source stations between adjacent receiver lines are recorded, the receiver patch is rolled over one line (or several lines), and the process is repeated. A portion of a 3-D layout is shown in Figure 2.1a, and a detailed view is presented in Figure 2.1b. In Chapters 2, 3, and 4, discussions concentrate on this layout method. Other methods that may be more suitable for solving particular problems are described in Chapter 5.

2.3 FOLD

Stacking fold (or fold-of-coverage) is the number of field traces that contribute to one stack trace, i.e., the number of midpoints per CMP bin. It is also the number of overlapping midpoint areas (see Section 5.2).

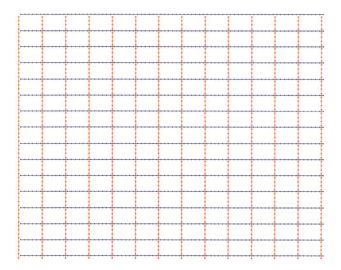

Fig. 2.1a. Orthogonal survey design.

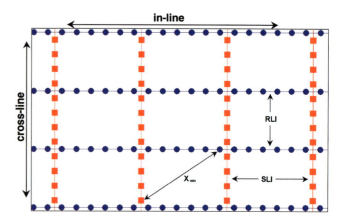

Fig. 2.1b. Orthogonal design—zoomed.

Fold controls the signal-to-noise ratio (S/N). If the fold is doubled, a 41% increase in S/N is accomplished (Figure 2.2). Doubling the S/N ratio requires quadrupling the fold, assuming that the noise is distributed in a random Gaussian fashion. Fold should be decided by looking at previous 2-D and 3-D surveys in the area, through evaluating X_{min} and X_{max} (Cordsen, 1995b), by modeling, and by remembering that dip moveout (DMO) and 3-D migration can effectively increase fold.

Krey (1987) showed that the ratio of 3-D to 2-D fold is frequency dependent and varies according to

$$\text{3-D fold} = \text{2-D fold} \times \text{frequency} \times C, \quad (2.1)$$

where C is an arbitrary constant.

For example, if $C = 0.01$ and 2-D fold = 40, then 3-D fold = 20 at 50 Hz and 40 at 100 Hz.

Rule of Thumb: Many designers use the equation 3-D Fold = 50% to 100% of 2-D Fold.

For example, if 2-D fold = 40, then a 3-D fold = $\frac{1}{2} \times 40 = 20$ usually achieves comparable signal-to-noise results to the 2-D data. To be on the safe side (especially if one expects high frequencies; e.g., over 100 Hz), one may define 3-D fold to be equal to the 2-D fold. Some designers recommend that 3-D fold be $\frac{1}{3} \times$ 2-D fold or even less. This lower ratio can give acceptable results only if the area has excellent S/N and only if there are minor problems with statics. The three-dimensional continuity of a 3-D data volume allows an easier correlation to neighboring lines than does 2-D data, hence a lower 3-D fold can be acceptable.

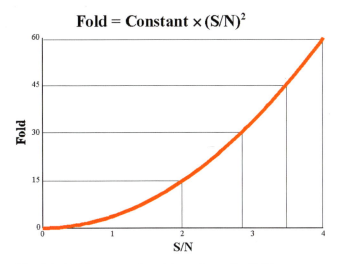

Fig. 2.2. Fold versus signal-to-noise ratio (S/N).

Krey's (1987) more complete formula for 3-D fold is

$$\text{3-D fold} = \text{2-D fold} \times \frac{\text{3-D bin spacing}^2 \times \text{frequency} \times \pi \times 0.401}{\text{2-D CDP spacing} \times \text{velocity}}. \quad (2.2)$$

As an example, if 2-D fold = 30 then,

$$\text{3-D fold} = 30 \times \frac{30^2 \text{ m}^2 \times 50 \text{ Hz} \times \pi \times 0.4}{20 \text{ m} \times 4500 \text{ m/s}} = 19$$

or

$$\text{3-D fold} = 30 \times \frac{110^2 \text{ ft}^2 \times 50 \text{ Hz} \times \pi \times 0.4}{55 \text{ ft} \times 15\,000 \text{ ft/s}} = 28.$$

If the 2-D trace spacing is much smaller than the 3-D bin size, then 3-D fold must be relatively higher to achieve results comparable to the 2-D imaging. However, large channel counts now mean that many 2-D surveys can be acquired with small trace spacing and large fold. Consequently many 2-D surveys are oversampled with higher than required fold. One must always keep this in mind when comparing 2-D and 3-D fold. In further support of a lower 3-D fold, one may consider trace (or sampling) density rather than geophone station density. Larger numbers of geophones per group certainly are sampling the subsurface more densely, and may improve data quality, when all 24 geophones are stacked into only one trace. However, 24 geophones per group do not necessarily provide better data than groups with 6 geophones. Similarly for sources, one may think of the sweep effort per square kilometer, where sweep effort is defined according to the description in Chapter 6 [particularly equation (6.2)].

There are many ways to calculate fold; the basic fact is that one source point creates as many midpoints as there are recording channels. If all offsets are within the acceptable recording range, then the basic fold equation is

$$\text{Fold} = SD \times NC \times B^2 \times U, \quad (2.3a)$$

where SD is the number of source points per unit area, NC is the number of channels, B is the bin dimension (for square bins), and

$U =$ units factor (10^{-6} for m/km²; 0.03587×10^{-6} for ft/mi²).

Derivation of equation (2.3a):

Number of midpoints = number of source points × NC

Source density $SD = \dfrac{\text{number of source points}}{\text{survey size}}$

Combine to obtain

$$\frac{\text{Number of midpoints}}{NC} = SD \times \text{survey size}$$

Survey size = Number of bins × bin size B^2.

Multiply with prior equation

$$\frac{\text{Number of midpoints}}{\text{Number of bins}} = SD \times NC \times B^2$$

$$\text{Fold} = SD \times NC \times B^2 \times U.$$

Example: Assuming that SD is 46/km² *(96/mi²)*, the number of channels NC is 720, and the bin dimension B is 30 m *(110 ft)* then,

$$\text{Fold} = 46 \times 720 \times 30 \times 30 \text{ m}^2/\text{km}^2 \times U$$
$$= 30{,}000{,}000 \times 10^{-6} = 30$$

or

$$\textit{Fold} = 96 \times 720 \times 110 \times 110 \text{ ft}^2/\text{mi}^2 \times U$$
$$= 836{,}352{,}000 \times 0.03587 \times 10^{-6} = 30.$$

16   Planning and Design

This formula is a quick way to calculate the average fold. To determine fold adequacy in a more detailed manner, one needs to examine the different components of fold. The following examples assume that the chosen bin size is small enough to satisfy the aliasing criteria.

In an orthogonal geometry the maximum in-line and cross-line offsets along with the receiver and source line intervals define the stacking fold fully. Different choices of station spacings will not influence fold, but will change the bin size, the source density, and the number of channels required.

A different way of solving equation (2.3a) is to solve for the number of channels NC. Once a certain fold requirement and bin size, source station and line intervals have been determined, the number of channels can be calculated as follows:

$$NC = Fold \div (SD \times B^2 \times U)$$
$$= Fold \times SLI \times SI \div B^2. \quad (2.3b)$$

2.4   IN-LINE FOLD

For an orthogonal straight-line survey, in-line fold is defined similarly to the fold on 2-D data. The formula is as follows:

$$\text{in-line fold} = \frac{\text{number of receivers} \times \text{station interval}}{2 \times \text{source interval along the receiver line}},$$

or

$$\text{in-line fold} = \frac{\text{number of receivers} \times RI}{\text{in-line patch dimension}}, \quad (2.4)$$

because the source line interval defines how many source points occur along any receiver line. It is important to use (number of receivers) × (RI) in equation (2.4) to describe the midpoint area that is covered. All receivers are assumed to be within the maximum usable offset range in these formulas. Figure 2.3a shows a smooth in-line fold distribution based on the following acquisition parameters [with one receiver line live (shown in blue) over many source lines]:

receiver interval	60 m	220 ft
receiver line interval	360 m	1320 ft
receiver line length	4320 m	15 840 ft (within the patch)

source interval	60 m	220 ft
source line interval	360 m	1320 ft

patch = 10 lines of 72 receivers.

Therefore,

$$\text{in-line fold} = \frac{4320 \text{ m}}{2 \times 360 \text{ m}} = 6$$

or

$$\text{in-line fold} = \frac{15\,840 \text{ ft}}{2 \times 1320 \text{ ft}} = 6.$$

If longer offsets are needed, care must be used in extending the in-line length. If a 9 × 80 patch was used instead of a 10 × 72 patch, the same number of channels (720) are employed, and the receiver line length is 80 × 60 m = 4800 m (80 × 220 ft = 17 600 ft). In this case,

$$\text{in-line fold} = \frac{4800 \text{ m}}{2 \times 360 \text{ m}} = 6.7$$

or

$$\text{in-line fold} = \frac{17\,600 \text{ ft}}{2 \times 1320 \text{ ft}} = 6.7$$

The offsets are indeed longer, but the in-line fold is now noninteger and shows striping as indicated in Figure 2.3b. Some of the fold values are 6 and some are 7, creating an average fold of 6.7 across the grid. This fold striping can be undesirable (a stack of 6

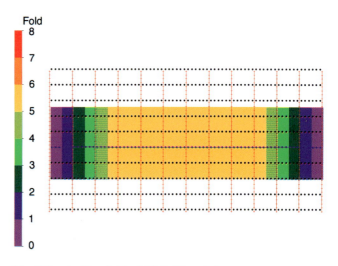

Fig. 2.3a. In-line fold of 10 × 72 patch.

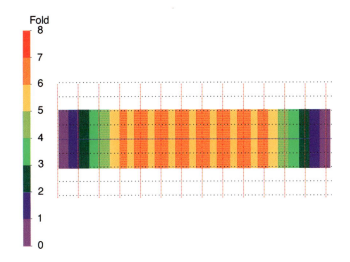

Fig. 2.3b. In-line fold of 9 × 80 patch.

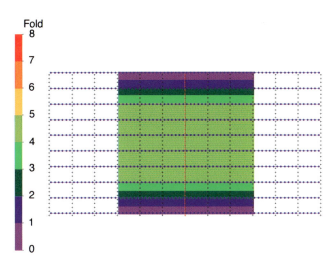

Fig. 2.4a. Cross-line fold of 10 × 72 patch.

traces may look quite different than a stack of 7 traces).

2.5 CROSS-LINE FOLD

Similar to the calculation of in-line fold, the cross-line fold is:

$$\text{cross-line fold} = \frac{\text{source line length}}{2 \times \text{receiver line interval}}$$

$$= \frac{\text{cross-line patch dimension}}{2 \times RLI} \quad (2.5)$$

Consequently,

$$\text{cross-line fold} = \frac{\text{number of receiver lines} \times RLI}{2 \times RLI} = \frac{NRL}{2}. \quad (2.6)$$

It is important to use (number of receiver lines) × (RLI) in equations (2.5) and (2.6) to define the midpoint area that is covered. Stated simply, the cross-line fold is half the number of live receiver lines in the recording patch (this is true for most geometries; exceptions will be discussed in Chapter 9).

In the original example of 10 receiver lines of 72 receivers each,

$$\text{cross-line fold} = \frac{10}{2} = 5.$$

Figure 2.4a shows such a cross-line fold by having just one source line live (shown in red) over ten receiver lines. Figure 2.4b, on the other hand, shows what happens when an odd number of receiver lines

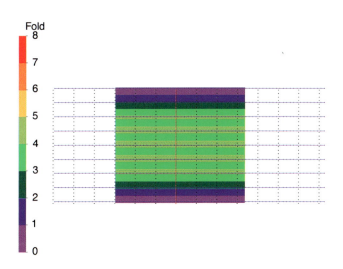

Fig. 2.4b. Cross-line fold of 9 × 80 patch.

are live within the patch (e.g., 9 lines). In this case, the cross-line fold varies between 4 and 5, because

$$\text{cross-line fold} = \frac{9}{2} = 4.5.$$

Generally this cross-line striping problem becomes less of a concern when one increases the number of receiver lines to, say 15, because the variation between 7 and 8 (15 ÷ 2 = 7.5) is much less on a percentage basis (12.5%) than the variation between 4 and 5 (20%). Nevertheless, anytime a patch is based on an odd number of receiver lines, some amount of cross-line striping will occur. Therefore most 3-D surveys are acquired with an even number of receiver lines live in the patch.

2.6 TOTAL FOLD

The total 3-D nominal fold is the product of in-line fold and cross-line fold:

total nominal fold =
  (in-line fold) × (cross-line fold). (2.7)

For the 10 × 72 patch example (Figure 2.5a), total nominal fold = 6 × 5 = 30. This value is the same value initially calculated using the formula:

$$\text{Fold} = SD \times NC \times B^2. \quad (2.8)$$

However, for the 9 × 80 patch, the in-line fold varies between 6 and 7 and the cross-line fold changes from 4 to 5; hence the total fold varies between 24 and 35 (Figure 2.5b). This 3-D fold oscillation is undesirable and results from lengthening the receiver lines. There were no changes in the source or receiver intervals or in the line intervals.

Note: The above equations assume that the bin size remains constant and is equal to half of the receiver interval, which in turn is equal to half the source interval. They also assume an orthogonal layout with all the source points within the patch.

By choosing the number of live receiver lines to be even, the cross-line fold is an integer and a smooth cross-line fold distribution results. Noninteger in-line and cross-line fold introduce striping in the 3-D fold distribution. If the maximum offset for stack exceeds the offset from any source point to any receiver station within the patch, then the smoothest fold distributions will result when the in-line and cross-line folds are integers (Cordsen, 1995b). Careful selection of the geometric configurations of the live patch is obviously one of the more significant components of 3-D design. The principles of design covered to this point can be summarized as

$$\text{in-line fold} = \frac{\text{in-line patch dimension}}{2 \times SLI} \quad (2.9)$$

and

$$\text{cross-line fold} = \frac{\text{cross-line patch dimension}}{2 \times RLI}. \quad (2.10)$$

These equations can be combined to yield

$$\text{total fold} = \text{in-line fold} \times \text{cross-line fold}, \quad (2.11)$$

or

$$\text{total fold} = \frac{\text{in-line patch dimension}}{2 \times SLI} \times \frac{\text{cross-line patch dimension}}{2 \times RLI}, \quad (2.12)$$

which is

$$\text{total fold} = \frac{\text{patch size}}{4 \times SLI \times RLI}. \quad (2.13)$$

Thus,

$$\text{total fold} = \frac{\text{patch size}}{4 \times \text{box size}}. \quad (2.14)$$

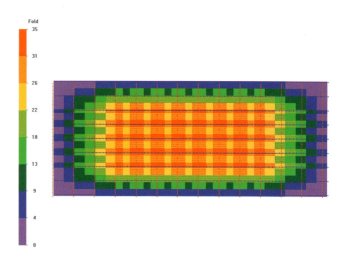

Fig. 2.5b. Total fold of 9 × 80 patch.

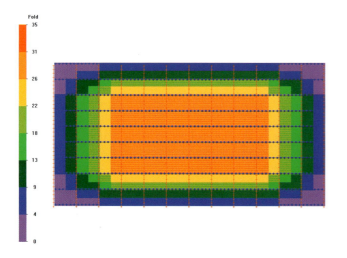

Fig. 2.5a. Total fold of 10 × 72 patch.

This formula holds true for rolling stations and lines on/off (see Chapter 9). Note that $1/4$ of the patch is the area in the subsurface that is covered by midpoints. Hence when rolling the receiver stations and lines, it is the quarter patches of midpoints that overlap to build up the fold.

Since the ratio of the area of a circle and a square patch is $\pi R^2 \div$ (patch size), the fold within a circle of radius R [compare equation (2.14)] is

$$\text{fold}_R = \frac{\pi R^2}{4 \times SLI \times RLI} = \frac{\pi R^2}{4 \times \text{box size}}, \quad (2.15)$$

as long as there is continuous coverage of receivers over the entire area and a square patch, which fits outside of the circle of radius R, is used. This equation estimates the fold for each horizon (depth) of interest as defined by that horizon's mute function (or X_{mute}). Equation (2.15) estimates the fold for circular patches as well. It is of note that this equation is totally independent of the station spacings; those merely define the natural bin size.

Goodway and Ragan (1995) compared 2-D fold and 3-D fold at a particular offset R. For 2-D data the fold is calculated as

$$\text{2-D fold}_R = \frac{\text{offset } R}{\text{source interval}}. \quad (2.16)$$

The ratio of 3-D and 2-D fold at offset R can then be defined as

Fold Ratio$_R$ (3-D/2-D) =
$$\frac{\pi R \times \text{2-D source interval}}{4 \times SLI \times RLI}. \quad (2.17)$$

This fold ratio is linear with offset R. Large line spacings (coarse sampling) lead to a low fold ratio, which might be acceptable for deeper targets (with the large increase in fold at far offsets). Decreasing the line spacings increases the fold ratio, therefore increasing fold at near offset, which is good for shallow targets. A compromise could be accomplished by using a narrow azimuth patch (see Section 3.3) to even out the fold distribution.

2.7 FOLD TAPER

Another important factor to consider when calculating fold is the fold taper. This parameter describes the area around the full-fold area where the fold build-up occurs. The width of this strip is not necessarily the same in the in-line and cross-line directions and needs to be calculated separately as follows:

$$\text{in-line taper} = \left(\frac{\text{in-line fold}}{2} - 0.5\right) \times SLI, \quad (2.18)$$

and

cross-line taper
$$= \left(\frac{\text{cross-line fold}}{2} - 0.5\right) \times RLI. \quad (2.19)$$

By substituting the formulas for in-line and cross-line fold, one can derive the following useful form of these equations:

$$\text{in-line taper} = \frac{\text{in-line patch size}}{4} - \frac{SLI}{2} \quad (2.20)$$

$$\text{cross-line taper} = \frac{\text{cross-line patch size}}{4} - \frac{RLI}{2} \quad (2.21)$$

Rule of Thumb: The fold taper is approximately equal to one quarter of the patch dimension in the fold-taper direction.

These equations define fold taper in units of meters *(or feet)*. A better way to express fold taper is in terms of source and receiver line intervals because that definition makes it easier to study the effects of fold taper when looking at fold maps. Hence the term "fold rate" is defined as the increase in fold per line interval in a specified direction, or

$$\text{in-line fold rate} = \frac{\text{total fold} \times SLI}{\text{in-line fold taper}}, \quad (2.22)$$

and

$$\text{cross-line fold rate} = \frac{\text{total fold} \times RLI}{\text{cross-line fold taper}}. \quad (2.23)$$

In the example of the 10 × 72 patch, the tapers and fold rates are as follows:

$$\text{in-line taper} = \left(\frac{6}{2} - 0.5\right) \times 360 \text{ m} = 900 \text{ m}$$

$$\text{cross-line taper} = \left(\frac{5}{2} - 0.5\right) \times 360 \text{ m} = 720 \text{ m}$$

20 *Planning and Design*

and

$$\text{in-line fold rate} = \frac{30 \text{ fold} \times 360 \text{ m}}{900 \text{ m}}$$
$$= 12 \text{ fold per } SLI,$$

which equals 2 ½ source line interval of fold taper, and

$$\text{cross-line fold rate} = \frac{30 \text{ fold} \times 360 \text{ m}}{720 \text{ m}}$$
$$= 15 \text{ fold per } RLI,$$

which equals 2 receiver line intervals.

2.8 SIGNAL-TO-NOISE RATIO (S/N)

For square bins, the S/N is directly proportional to the length of one side of the bin (Figure 2.6). Therefore, only a slight change in the selection of the bin size can have a major effect on the fold and the S/N. The designer of a 3-D survey needs to be given clear and precise specifications for these parameters to effectively optimize the 3-D design. If the fold drops below the required level for only a few bins, that does not necessarily mean that the 3-D survey is poorly designed. Increasing the fold by only a small percentage on an otherwise well-designed survey may cost an unreasonable amount of money to satisfy the fold requirements of a few bins.

2.9 BIN SIZE

It is important to differentiate between the bin size and the bin interval. The bin size is the area over which the traces are stacked. The bin interval determines how far apart these trace summations are displayed. Most of the time bin dimension and bin interval are used interchangeably (as they are in this text), because they have the same value, but occasionally they may differ (e.g. flex-binning in marine surveys).

The selection of bin size and fold go hand in hand. The fold is a quadratic function of the length of one side of the bin (Figure 2.7). The basic fold equation derived in Section 2.3 indicates that the constant relating fold to (bin size)^2 is the midpoint density (i.e., the number of midpoints per square unit area), or

$$\text{fold} = SD \times NC \times B^2. \quad (2.24)$$

The preferred shape of a 3-D data bin is a square. Rectangular bins may be acceptable to highlight certain geological features if the lateral resolution needed in one direction is different from the required resolution in the other direction. Also, the spatial sampling requirements for migration might be different in different directions. Sometimes cost issues will determine a different receiver station than source point interval; hence, natural bin sizes may differ. In some cases, rectangular bins may create problems because the smaller number of subsurface measurements in the long direction of the bins limits the resolving power of geological features in that direction.

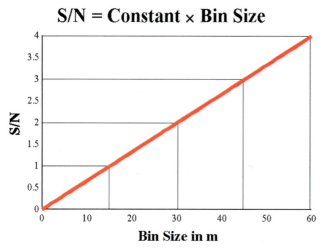

Fig. 2.6. Signal-to-noise ratio (S/N) versus bin size.

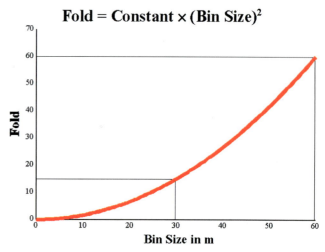

Fig. 2.7. Fold versus bin size.

Bin size can be determined by examining three factors: target size, maximum unaliased frequency due to dip, and lateral resolution, and then picking the smallest value of bin size provided by these analyses as the design parameter.

2.9.1 Target Size

Normally two to three traces, positioned so they pass through a small target, will allow that target to be seen in a 3-D image, because this means four to nine traces will be related to the target on a time slice of the horizon of interest. For example, if the target is a small reef or a narrow channel sand then the bins should be small enough to get at least two (preferably three) traces across the target. This imaging requirement gives a 3-D designer an initial (and generally too large) estimate for a bin size, which is

$$\text{Bin size} \leq \frac{\text{target size}}{3}. \tag{2.25}$$

Consider the following example where a 3-D survey in Alberta crossed a 300-m ($\frac{1}{5}$ mi) wide channel (Cordsen, 1993b), which had been difficult to define with 2-D data. Within this narrow channel, a 100-m ($\frac{1}{16}$ mi) wide sand anomaly surrounded by shale could be identified on the 3-D data (Figure 2.8). The choice of 24 m × 24 m (78 ft × 78 ft) bin size allowed the sand anomaly to be recognized on four traces crossing the channel. This 4-trace response is close to the minimum that is required for target recognition by many interpreters. Had the bin size been chosen much larger, the sand anomaly may not have shown up at all.

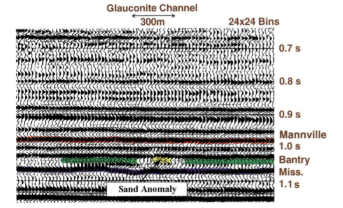

Fig. 2.8. Bin size and target size.

The power of 3-D data is that an interpreter can relate anomalies from one cross-line to the next cross-line and follow the seismic expression continuously. Older 2-D data would not have convinced management to drill this narrow sand body, but several successful oil wells have been drilled into this channel based on the 3-D data.

2.9.2 Maximum Unaliased Frequency

Each dipping seismic reflection event has a maximum possible unaliased frequency f before migration that depends on the velocity to the target, the value of the geological dip θ, and the bin size B. Referring to Figure 2.9a, these parameters are related as

$$\sin \theta = \frac{V \times \Delta t}{B}. \tag{2.26}$$

One needs to take account of the fact that Δt represents only $\frac{1}{4}$ wavelength since two-way traveltime is measured and two samples per wavelength are required to avoid aliasing. Thus

$$\Delta t = \frac{\lambda}{4 \times V} = \frac{1}{4 \times f}, \tag{2.27}$$

and replacing Δt

$$\sin \theta = \frac{V}{4 \times B \times f}. \tag{2.28}$$

Therefore,

$$f = \frac{V}{4 \times B \times \sin \theta} \tag{2.29}$$

and

$$B = \frac{V}{4 \times f \times \sin \theta}. \tag{2.30}$$

The reflector dip θ is very important in these two equations. A negligible dip produces very large values for the largest bin size, which does not cause aliasing, and for maximum unaliased frequency. The largest dip of 90° puts the most constraint on these calculations. The main question is to decide which velocities or frequencies should be used for the bin size calculations. Common practice has been to use the average velocity V_{ave} and the dominant frequency f_{dom}

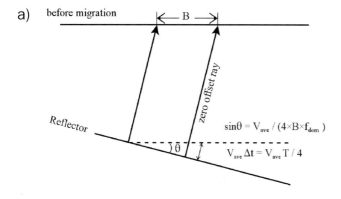

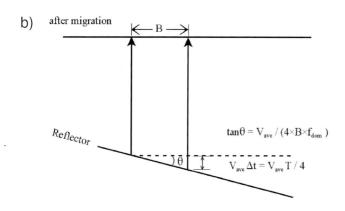

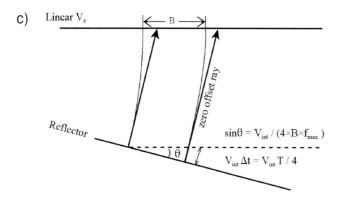

Fig. 2.9. Bin size B and maximum unaliased frequency; a. before migration, b. after migration, c. linear-velocity earth.

for a constant-velocity earth (as in Figures 2.9a and 2.9b), giving

$$f_{dom} = \frac{V_{ave}}{4 \times B \times \sin\theta}. \quad (2.31)$$

For example:

$$f_{dom} = \frac{2500 \text{ m/s}}{4 \times 25 \text{ m} \times \sin 15°} = 97 \text{ Hz}$$

or

$$f_{dom} = \frac{7500 \text{ ft/s}}{4 \times 82.5 \text{ ft} \times \sin 15°} = 88 \text{ Hz}.$$

Solving for the bin size B,

$$B = \frac{V_{ave}}{4 \times f_{dom} \times \sin\theta}, \quad (2.32)$$

yields bin size values of (if $f_{dom} = 60$ Hz):

$$B = \frac{2500 \text{ m/s}}{4 \times 60 \text{ Hz} \times \sin 15°} = 40 \text{ m},$$

or

$$B = \frac{7500 \text{ ft/s}}{4 \times 60 \text{ Hz} \times \sin 15°} = 121 \text{ ft}.$$

Most geological scenarios do not warrant the constant-velocity medium assumption. A velocity that increases linearly with depth is a better assumption in many basins. A common velocity function is

$$V_z = V_0 + kZ, \quad (2.33)$$

where V_z is the depth-varying velocity, V_0 is the velocity at surface, k is a constant (usually >0), and Z is depth. Margrave (1997) used this depth-varying velocity to determine the bin size.

One needs to consider ray-bending to avoid over-constraining the bin size (see Bee et al., 1994). An example of ray-bending is illustrated in Figure 2.9c (after Liner and Gobeli, 1997). The raypaths are parallel for ray parameter p until the up-dip raypath reaches the reflector. The ray parameter p is a constant that is independent of depth and is defined as

$$p = \frac{\sin\theta_0}{V_z}, \quad (2.34)$$

where θ_0 is the take-off angle for the ray rather than the geological dip. The bin size for a depth varying velocity model can be calculated as follows:

$$B = \frac{V_z}{4 \times f_{max} \times \sin\theta}. \quad (2.35)$$

The interval velocity V_{int} immediately above the horizon [or V_z in equation (2.35)], rather than the average velocity, should be used for calculations of bin size at the target. This choice of bin size assures that the maximum frequency at the target f_{max} is not aliased with reflector dip θ. Therefore,

$$f_{max} = \frac{V_{int}}{4 \times B \times \sin \theta}. \quad (2.36)$$

For example:

$$f_{max} = \frac{3000 \text{ m/s}}{4 \times 25 \text{ m} \times \sin 15°} = 116 \text{ Hz}$$

or

$$f_{max} = \frac{10\,000 \text{ ft/s}}{4 \times 82.5 \text{ ft} \times \sin 15°} = 118 \text{ Hz}.$$

Solving for the bin size B,

$$B = \frac{V_{int}}{4 \times f_{max} \times \sin \theta}. \quad (2.37)$$

For example if $f_{max} = 80$ Hz, then

$$\text{bin size } B = \frac{3000 \text{ m/s}}{4 \times 80 \text{ Hz} \times \sin 15°} = 36 \text{ m}$$

or

$$\text{bin size } B = \frac{10\,000 \text{ ft/s}}{4 \times 80 \text{ Hz} \times \sin 15°} = 120 \text{ ft}.$$

Any frequencies at the zone of interest that are higher than f_{max} will be aliased before migration. In other words, the true dip of the event will be contained only in frequencies lower than this value. For the following discussions, the maximum frequency and the interval velocity at the target are used.

The above equations are based on recording two samples per wavelength of the maximum frequency. Many companies use more stringent requirements of three or four (or even noninteger values such as 2.8) samples per wavelength of the dominant frequency, which greatly reduces the bin size and increases the survey cost.

Note that the process of migration lowers frequencies on all dipping events; the rule being that the steeper the dip, the lower the frequencies after migration. Any aliasing of frequencies prior to migration may look like frequency dispersion after migration due to the particular migration algorithm that is used. The correct choice of bin size successfully preserves the desired maximum frequency through the migration step. The connection between bin size B and f_{max} after migration is given by similar formulas as above in which $\sin \theta$ is replaced by $\tan \theta$ (Figure 2.9b).

Metric Example:

Using the formula, $f_{max} = \dfrac{V_{int}}{4 \times B \times \sin \theta}$, calculate f_{max} for the following three sets of numbers ($V_{int} = 3000$ m/s):

Dip (in degrees)	B (meters)	f_{max} (Hz)
5	100	
20	25	
35	15	

Using the formula, $B = \dfrac{V_{int}}{4 \times f_{max} \times \sin \theta}$, calculate the bin size B for the following three sets of numbers ($V_{int} = 3000$ m/s):

Dip (in degrees)	f_{max} (Hz)	B (meters)
15	100	
25	60	
40	40	

Imperial Example:

Using the formula, $f_{max} = \dfrac{V_{int}}{4 \times B \times \sin \theta}$, calculate f_{max} for the following three sets of numbers ($V_{int} = 10\,000$ ft/s):

Dip (in degrees)	B (ft)	f_{max} (Hz)
5	440	
20	110	
45	55	

Using the formula, $B = \dfrac{V_{int}}{4 \times f_{max} \times \sin \theta}$, calculate the bin size B for the following three sets of numbers ($V_{int} = 10\,000$ ft/s):

Dip (in degrees)	f_{max} (Hz)	B (ft)
15	80	
20	60	
30	40	

Alias Frequency Tables
$V_\text{int} = 3000$ m/s

Table 2.2. Calculation of f_max.

$$f_\text{max} = \frac{V_\text{int}}{4 \times \text{bin size} \times \sin\theta}$$

Dip (degrees)	Subsurface trace spacing in meters						
	10	15	20	25	30	50	100
5	861	574	430	344	287	172	**86**
10	432	288	216	173	144	86	43
15	290	193	145	116	97	58	29
20	219	146	110	**88**	73	44	22
25	177	118	89	71	59	35	18
30	150	100	75	60	50	30	15
35	131	**87**	65	52	44	26	13
40	117	78	58	47	39	23	12
45	106	71	53	42	35	21	11	in Hz

Table 2.3. Calculation of bin size B.

$$B = \frac{V_\text{int}}{4 \times f_\text{max} \times \sin\theta}$$

Dip (degrees)	f_max in Hz					
	40	50	60	70	80	90	100
5	215	172	143	123	108	96	86
10	108	86	72	62	54	48	43
15	72	58	48	41	36	32	**29**
20	55	44	37	31	27	24	22
25	44	35	**30**	25	22	20	18
30	38	30	25	21	19	17	15
35	33	26	22	19	16	15	13
40	**29**	23	19	17	15	13	12
45	27	21	18	15	13	12	11	in meters

Alias Frequency Tables
$V_\text{int} = 10\,000$ ft/s

Table 2.4. Calculation of f_max.

$$f_\text{max} = \frac{V_\text{int}}{4 \times B \times \sin\theta}$$

Dip (degrees)	Subsurface trace spacing in feet						
	41	55	82.5	110	165	220	440
5	700	522	348	261	174	130	**65**
10	351	262	175	131	87	65	33
15	236	176	117	88	59	44	22
20	178	133	89	**66**	44	33	17
25	144	108	72	54	36	27	13
30	122	91	61	45	30	23	11
35	106	79	53	40	26	20	10
40	95	71	47	35	24	18	9
45	86	**64**	43	32	21	16	8	in Hz

Table 2.5. Calculation of bin size B.

$$B = \frac{V_\text{int}}{4 \times f_\text{max} \times \sin\theta}$$

Dip (degrees)	f_max in Hz					
	40	50	60	70	80	90	100
5	717	574	478	410	359	319	287
10	360	288	240	206	180	160	144
15	241	193	161	138	**121**	107	97
20	183	146	**122**	104	91	81	73
25	148	118	99	85	74	66	59
30	**125**	100	83	71	63	56	50
35	109	87	73	62	54	48	44
40	97	78	65	56	49	43	39
45	88	71	59	51	44	39	35	in feet

Such aliasing frequency tables could easily be developed for other interval velocities V_int relative to the area of interest.

2.9.3 Lateral Resolution

Before migration, two diffractions are not resolved if they are closer than the first Fresnel zone diameter. This diameter is usually a large value (500 m or more) and means that on the CMP stack, small, closely spaced faults can be missed. After migration, the lateral resolution depends on the maximum frequency that is reflected from the zone of interest. In the past several years, different definitions and formulas for lateral resolution have been published.

Denham (1980):

$$R_H = \frac{R_V}{\sin\theta},$$

where θ is the maximum ray angle contributing to the migration, R_H is horizontal resolution, and R_V is vertical resolution.

Claerbout (1985):

"Resolving power is customarily defined as about half the effective wavelength. . . ."

Embree (1985):

$$dX_r = \frac{0.3 \times V}{f_\text{max} \times \sin\theta}.$$

Freeland and Hogg (1990):

$$R_L = R_V = \frac{V}{2B},$$

where R_L = lateral resolution, R_V = vertical resolution, and B = bandwidth. This equation assumes time period = half bandwidth.

Ebrom et al. (1995):

$$R_L = \frac{\lambda_{min}}{4 \times \sin \theta},$$

where $\lambda_{min} = \frac{1}{f_{max}}$, $\theta = \arctan \frac{L}{2Z}$, $\frac{L}{2}$ is half the line length, and Z is the target depth.

Vermeer (1997):

$$R_x = \frac{V}{4 \times f_{max} \times \sin \theta_{x, max}}.$$

Vermeer interprets this as $\frac{V_{min}}{4 \times f_{max}}$, where V_{min} is the minimum apparent velocity of the P-wave data and $\theta_{x, max}$ is the scattering angle and not reflector dip.

Finite recording apertures (patches) in space and time imprint a strong spatial-temporal variation on the spectral content of the migrated section (Margrave, 1997). Not everyone agrees on the definition of resolution: Is it one half the dominant wavelength (Claerbout, 1985; Freeland and Hogg, 1990), or is it one quarter of the minimum wavelength (Ebrom, et al., 1995; Vermeer, 1997), or somewhere in between (Embree, 1985)? One of the key questions regarding resolution is the ability of the interpreter to recognize the feature being resolved. This ability is interpreter-dependent and should be considered when picking the bin size.

In 3-D surveys, the migration aperture varies throughout the survey because of edge effects. It is therefore reasonable to use a migration aperture angle such as 30° in Vermeer's formula. It is also reasonable to expect the resolution to change by a factor of 2 or more from the edges of the survey to the center. The existence of noise (ground roll, multiples, etc.) causes further loss of resolution to a limit that is commonly termed "achievable resolution." In practice, the "actual resolution" will be less than the "achievable resolution" because of errors in processing such as using the wrong velocity model, suboptimal phase compensation, etc.

For the purposes of the present discussion, it is assumed that lateral resolution will be between one-quarter and one-half the dominant wavelength. The dominant frequency can be measured directly from the seismic section, thus a simple guideline equation for bin size is:

$$\text{Bin Size } B = \frac{V_{int}}{N \times f_{dom}}, \quad (2.38)$$

where N varies from 2 to 4. Suppose the dominant frequency f_{dom} at the selected target is 50 Hz. If the interval velocity immediately above this target is 3000 m/s (10 000 ft/s), then the spatial wavelength,

λ_{dom} is $\frac{3000 \text{ m/s}}{50 \text{ Hz}} = 60 \text{ m}$ or $\left(\frac{10\,000 \text{ ft/s}}{50 \text{ Hz}} = 200 \text{ ft}\right)$,

and the bin size is:

bin size $B = \frac{3000 \text{ m/s}}{2 \times 50 \text{ Hz}} = 30$ m (or 15 m for $N = 4$),

or

bin size $B = \frac{10\,000 \text{ ft/s}}{2 \times 50 \text{ Hz}} = 100$ ft (or 50 ft for $N = 4$).

Using the examples of the last few pages, the lateral resolution defines the range of possible bin sizes from a minimum of 15 m (50 ft) to a maximum acceptable bin size of 30 m (100 ft) (Table 2.6). There is little point in choosing a bin size less than that needed for lateral resolution. Smaller bins would provide no additional information. Correspondingly, with larger bins there is a danger that some events will not be resolved laterally.

A bin size less than one quarter of the dominant wavelength results in over-sampling and provides no additional information. A bin size greater than half of the dominant wavelength results in spatial aliasing and missing information.

Table 2.6. Selection of bin size.

Parameter	Bin Size
Target Size	33 m *(100 ft)*
Maximum Unaliased Frequency	36 m *(120 ft)*
Lateral Resolution	15–30 m *(50–100 ft)*

A bin size of 30 m is selected for this example.

2.9.3.1 Lateral Resolution after Migration

The lateral resolution that exists after migration can be explained using Figure 2.10. Four events are shown in f-k space both before and after migration. The effect of migration is to map high frequencies to low frequencies at the same wavenumber. The

26 *Planning and Design*

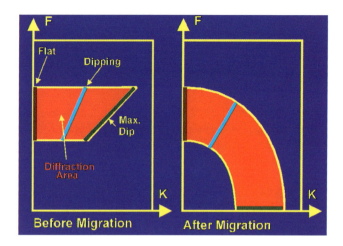

Fig. 2.10. Lateral resolution before and after migration.

amount of movement increases with increasing wavenumber (k). This effect is true for any migration algorithm. Key criteria involved in migration are the following:

Flat event: The energy for all flat events lies along the frequency axis and is confined to the passband of the reflection signal. Flat events are unaffected by migration.

Dipping event: With migration, intermediate dipping events are converted to a steeper dip, experience a general lowering of frequency, and have less bandwidth.

Maximum dip: This event has the maximum dip before migration that will appear as a dip of 90° after migration. The event has no frequency after migration and appears as a dc bias in the migrated data.

Diffraction: Before migration, a diffraction occupies the entire colored zone with limited bandwidth. After migration, the diffraction is collapsed and occupies the entire colored zone in the right part of Figure 2.10. Migration maps the *f-k* points to other points at lower frequencies. The diffraction now has frequencies from zero to the maximum frequency before migration and wavenumbers from zero to the maximum spatial frequency. The diffraction is converted into a bandlimited spike, both in frequency and in wavenumber. Its temporal resolution (equal to one wavelength of the maximum frequency) is equal to the spatial resolution (one wavelength of the maximum frequency expressed in spatial units).

2.9.3.2 Separation of Diffractions

The following four synthetic displays show two diffractors separated by two intermediate traces with 10-m *(33 ft)* trace spacing and a velocity of 3000 m/s *(10 000 ft/s)*; thus the diffraction points are 30 m *(100 ft)* apart.

In the first pair of figures (Figure 2.11a before migration, Figure 2.11b after migration), the maximum frequency is set to 100 Hz, and the two events are well resolved after migration. In fact, this case is the limit of spatial resolution for the given frequency and event separation. The spatial wavelength is 30 m *(100 ft)*,

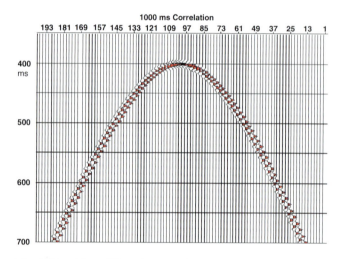

Fig. 2.11a. Two diffractions before migration, 100 Hz wavelet, 30 m *(100-ft)* lateral separation.

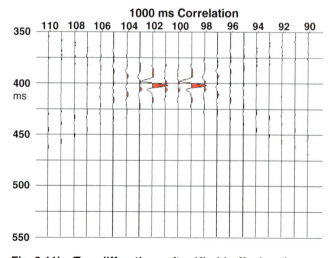

Fig. 2.11b. Two diffractions after Kirchhoff migration, 100 Hz wavelet.

which is borne out by the facts that each diffractor after migration is on a peak and that there are two traces (with troughs) between the events. Wavelength values can be calculated using

$$V_{int} = f_{max} \times \lambda_{max}, \quad (2.39)$$

or

$$\lambda_{max} = \frac{V_{int}}{f_{max}}, \quad (2.40)$$

or

$$\lambda_{dom} = \frac{V_{int}}{f_{dom}} \cong 2\lambda_{max}, \quad (2.41)$$

where V_{int} is the velocity immediately above target, f_{max} is maximum frequency, f_{dom} is the dominant frequency $\cong \frac{1}{2} f_{max}$, λ_{max} is the wavelength at f_{max}, and λ_{dom} is the wavelength at f_{dom}.

In this example:

$$\text{spatial wavelength} = \frac{3000 \text{ m/s}}{100 \text{ Hz}} = 30 \text{ m}$$

or

$$\textit{spatial wavelength} = \frac{10\,000\,\textit{ft/s}}{100\,\textit{Hz}} = 100\,\textit{ft}.$$

The spatial resolution is also 30 m (100 ft), i.e., one wavelength of the maximum frequency, or one-half-wavelength of the dominant frequency. In this case the maximum frequency was used in the calculation. One can get the same answer by using the half-wavelength of the dominant frequency of 50 Hz.

The second pair of figures (Figure 2.12a before migration, Figure 2.12b after migration) shows the effect of reducing the maximum model frequency to 50 Hz, which corresponds to a spatial wavelength of 60 m (200 ft). In this case, both diffractors are on the same positive amplitude "spatial wavelet" and the lateral resolution is lost.

$$\text{Spatial wavelength} = \frac{3000 \text{ m/s}}{50 \text{ Hz}} = 60 \text{ m}$$

or

$$\textit{Spatial wavelength} = \frac{10\,000\,\textit{ft/s}}{50\,\textit{Hz}} = 200\,\textit{ft}.$$

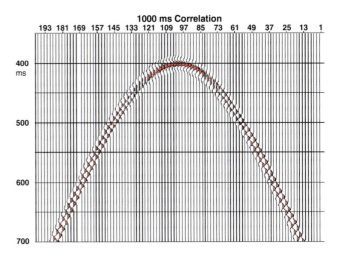

Fig. 2.12a. Two diffractions before migration, 50 Hz, 30 m *(100-ft)* lateral separation.

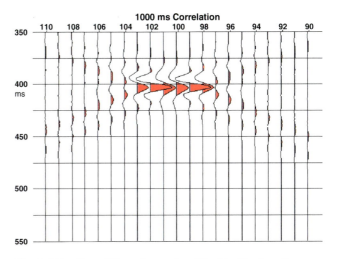

Fig. 2.12b. Two diffractions after Kirchhoff migration, 50 Hz wavelet.

2.9.4 Vertical Resolution

A classic paper on vertical (temporal) resolution has been published by Kallweit and Wood (1982). They discuss various criteria (Rayleigh, Ricker, and Widess) and how their criteria can be used to describe the width of a wavelet as a measure of temporal resolution. In Vermeer (1999a), it is pointed out that the configuration of 3-D sources and receivers leads to a significant difference between vertical and lateral resolution. Vermeer suggests the following formula as a more practical alternative to the more optimistic formula derived from zero-offset considerations and Widess criteria ($R_z = \frac{1}{4} V/f_{max}$),

$$R_z = cV/(2 f_{max} \cos i), \quad (2.42)$$

where R_z is the vertical resolution in spatial units, c is a constant depending on the criterion used (Rayleigh, etc.), V is the local interval velocity over the physical dimension of a temporal wavelet, f_{max} is the maximum frequency recorded from the target, and cos i is the cosine of angle i that is the half-angle subtended between source and receiver to a point at depth; cos i can be interpreted as the NMO stretch which reduces f_{max} to $f_{max} \times \cos i$.

Example:

$$V = 2500 \text{ m/s},$$
$$c = 0.715 \text{ (Rayleigh criterion)},$$
$$\cos i = 0.9, f_{max} = 40 \text{ HZ}.$$

Then

$$R_z = 0.715 \times 2500/2 \times 40 \times 0.9 = 25 \text{ m}.$$

In most cases, it is best to use $c = 0.715$. In some cases, it may be possible to use $c = 0.25$ (1/4 wavelength criterion) if all parameters around a target horizon are more or less constant except for thickness. In such a case, tuning effects can be used with confidence. The value of cos i can also be set to 0.9, which corresponds approximately to the "maximum offset equals depth criterion" for muting (NMO stretch).

Thus a practical equation is

$$R_z = 0.715 \, V/(2 f_{max} \, 0.9) = 0.4 \, V/f_{max}. \quad (2.43)$$

This equation states that the vertical resolution R_z is $\frac{2}{5}$ of the wavelength of the maximum frequency reflecting from the target to be resolved.

Let's Design a 3-D—Part 1

For the following exercise a simple approximation for calculating the sine can be used, which is

$$\sin(\text{dip}) \cong \frac{\text{dip}}{60}.$$

For example:

$$\sin 6° \cong 0.1$$
$$\sin 12° \cong 0.2$$
$$\sin 18° \cong 0.3$$

This approximation holds true within 6% for angles <45°.

In this example, these are the following known parameters (metric):

existing 2-D data of good quality have	30 fold
steepest dips	20°
shallow markers needed for isochroning have	500 m offsets
target depth	2000 m
target two-way time	1.5 s
basement depth	3000 m
V_{int} immediately above the target horizon	4200 m/s
f_{dom} at the target horizon	50 Hz
f_{max} at the target horizon	70 Hz
lateral target size	300 m

Orthogonal design method should be used.

Desired fold:

$$\tfrac{1}{2} \text{ up to full 2-D fold} =$$

Bin size:

a) for target size: B = target size ÷ 3 =
b) for alias frequency: $B = V_{int} \div (4 \times f_{max} \times \sin \theta) =$
c) for lateral resolution: $B = V_{int} \div (N \times f_{dom})$, ($N$ = 2 to 4) = ____ to ____
bin size B =
RI =
SI =

Let's Design a 3-D—Part 1 (Imperial)

In this example, these are the following known parameters:

existing 2-D data of good quality have	30 fold
steepest dips	20°
shallow markers needed for isochroning have	1500 ft offsets
target depth	6000 ft
target two-way time	1.5 s
basement depth	10 000 ft
V_{int} immediately above the target horizon	14 000 ft/s
f_{dom} at the target horizon	50 Hz
f_{max} at the target horizon	70 Hz
lateral target size	1000 ft

Orthogonal design method should be used.

Desired fold: $\frac{1}{2}$ up to full 2-D fold =

Bin size:

  a) for target size: B = target size ÷ 3 =
  b) for alias frequency: B = V_{int} ÷ (4 × f_{max} × sin θ) =
  c) for lateral resolution: B = V_{int} ÷ (N × f_{dom}), (N = 2 to 4) = ____ to ____
  bin size B =
  RI =
  SI =

2.10 X_{MIN}

The bin at the center of the box formed by two adjacent receiver lines and two source adjacent lines has the largest minimum offset of any bin within the box. The largest minimum offset therefore is the diagonal of this box (Figure 2.13a). The source and receiver line intervals (*SLI* and *RLI*) are largely determined by the required value for X_{min}. In orthogonal, brick, and zig-zag designs (see Chapter 5), the largest minimum offset is related directly to *SLI* and *RLI*. For orthogonal surveys, X_{min} is determined by:

$$X_{min} = (RLI^2 + SLI^2)^{1/2}. \qquad (2.44)$$

Obviously, the larger of *RLI* and *SLI* has the greater influence on X_{min}. Therefore, in an orthogonal design, a square box is the ideal case for minimizing X_{min}.

Going back to the practice example:

$$X_{min} = (360^2 + 360^2)^{1/2} \text{ m} = 509 \text{ m}$$

or

$$X_{min} = (1320^2 + 1320^2)^{1/2} \text{ ft} = 1867 \text{ ft}.$$

Generally lines are offset from their coincident positions by one bin size at their intersection points (Figure 2.13b) in order to reduce duplicate raypaths. The formula for X_{min} changes slightly for this geometry as follows:

$$X_{min} = [(RLI - 0.5 \times SI)^2 + (SLI - 0.5 \times RI)^2]^{1/2}. \qquad (2.45)$$

When offsetting the source or receiver lines in this way, the four bins in the center of the box have the same X_{min}, rather than just the center bin having the largest X_{min}. Going back to the practice example:

$$X_{min} = [(360 - 30)^2 + (360 - 30)^2]^{1/2} \text{ m} = 467 \text{ m}$$

or

$$X_{min} = [(1320 - 55)2 + (1320 - 55)^2]^{1/2} \text{ ft} = 1789 \text{ ft}.$$

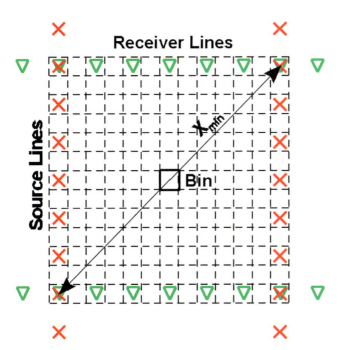

Fig. 2.13a. X_{min} definition with coincident source and receiver stations at corners of box.

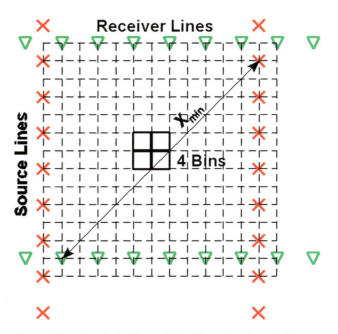

Fig. 2.13b. X_{min} definition with half-station line shifts at box corners.

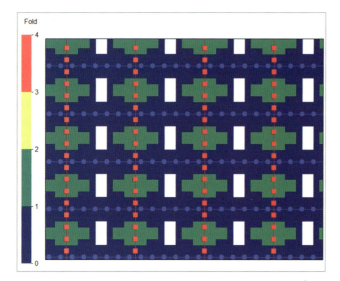

Fig. 2.14. Fold distribution at a shallow level for an X_{min} that is too large.

X_{min} should be small enough to adequately sample shallow reflectors that might be used for datuming or isochroning purposes. If such a shallow marker is not sampled appropriately (because the mute is too tight), the interpretability of the 3-D data set is adversely affected.

Figure 2.14 shows what happens when a configuration of source and receiver line intervals is chosen which creates an X_{min} that is too large for the shallow reflector criterion. In the center of the boxes, holes develop in the fold distribution. Insufficient fold may image a shallow marker in an inconsistent and unreliable manner, and picking horizon times or amplitudes might be impossible. This inability to pick shallow reflection times in both cases reduces the reliability of any interpretation, flattening, or mapping based on shallow markers.

Single-fold stacks in the bins that have the largest X_{min} are often insufficient for accurate interpretation, or for picking horizon times for isochroning from the shallow horizon. A schematic seismic section along the direction of the X_{min} measurement (Figure 2.15a) shows the lack of data at early arrival times. These data gaps are caused by the mute pattern indicated in Figure 2.15b where single-fold data are assumed below t_{min}. Usually at least four-fold multiplicity is necessary to have enough confidence in a correct interpretation at a shallow datuming marker Z_{sh} (at time t_{sh}). Vermeer (1998a) has provided such a four-fold formula for symmetric sampling at a mute distance X_{sh} (Figure 2.15c) given by

$$X_{sh} = RLI \times 2 \times \sqrt{2} \\ = SLI \times 2 \times \sqrt{2} = 2 \times X_{min}. \quad (2.46)$$

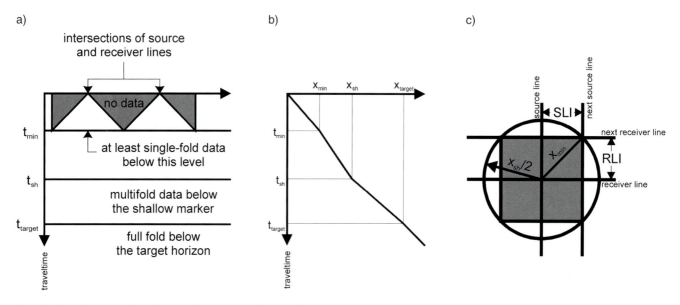

Fig. 2.15a. A *t-x* section diagonally through the middle of a box, b. Typical mute function showing the relationship between X_{min}, X_{sh}, and X_{target}, c. Required relationship between X_{min} and X_{sh} to achieve 4-fold data at X_{sh} (after Vermeer, 1998a).

2.10 X_{MIN} 31

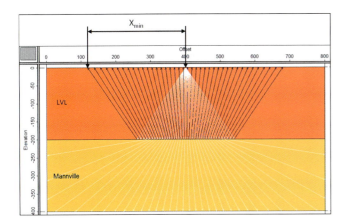

Fig. 2.16. Modeling X_{min}.

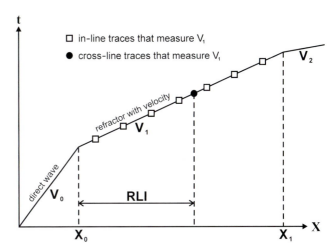

Fig. 2.17a. Refraction criterion—t-x diagram.

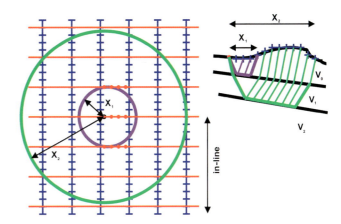

Fig. 2.17b. Refraction criterion—plan view and section view.

Equation 2.46 is somewhat pessimistic and does not account for source and receiver line spacings that are different. An improved estimate is provided by the following equation, which includes contributions from both line spacings. This estimate is generally good within 10% of the offsets required to reach four fold.

$$X_{sh} = \text{Max}(RLI, SLI) + \text{Min}(RLI, SLI) \times \sqrt{2} \quad (2.47)$$

A geological model such as the one in Figure 2.16 helps to determine the offsets at which critical refraction occurs. If one considers the shallowest horizon to be mapped, reflections will occur out to this critical refraction angle. Typically (and neglecting dip), critical refraction occurs at an incidence angle of about 35°, with a corresponding offset of $2 \times Z_{sh} \times \tan 35° = 1.4 \times Z_{sh}$, where Z_{sh} is the depth of the shallowest horizon to be mapped. Meeting this criterion by selecting an X_{min} which is in the range 1.0 to $1.2 \times Z_{sh}$ guarantees single-fold data, which probably is inadequate for mapping. If the X_{min} for the survey is larger than determined by this model (Figure 2.16; e.g., 400 m), then there will be holes in the reflected energy that is stacked. These data gaps lead to a fold distribution similar to the one indicated in Figure 2.14. The mute function (from 2-D data) can often be the best guide to this choice of X_{min}.

Rule of Thumb: X_{min} should be less than 1.0 to 1.2 Z_{sh}.

The refraction criterion says that there should always be at least three measurements of the shallow refractor to adequately sample the refractor velocity V_1 (Figure 2.17). The in-line direction provides a large number of refraction measurements of course. The first cross-line measurement of V_1 must be within a receiver line interval of the first in-line measurement of V_1. Failure to provide common receivers in the cross-line direction for a shallow refractor means that the refraction statics solutions are not fully coupled from one receiver line to the next. Thus static shifts from one receiver line to the next are indeterminate. Existing 2-D data can provide critical information about the first refractor. Since the receiver line interval affects X_{min}, RLI should be chosen small enough to guarantee the necessary refraction measurements.

An alternative to satisfying the refraction criterion is to record separate 2-D data along each source and receiver line. This will completely determine source and receiver delays (statics) and near-surface velocities. The cost of a separate 2-D refraction statics survey may well be justified for 3-D surveys with widely spaced source and receiver lines rather than reducing the line intervals.

2.11 X_{MAX}

The required maximum offset depends on the depth to the deeper targets that must be imaged. One also needs to take into account normal moveout (NMO) assumptions and dip. It is strongly suggested that the designer pay particular attention to the offset distribution on a line-by-line basis.

The processing mute of the far offsets has a great impact on selection of the maximum recorded offset (Cordsen, 1995b). If X_{max} in the patch is measured as the maximum in-line offset (Figure 2.18a), then traces on receiver lines farther away from the source point are muted. The fold value is limited by X_{mute}. Since the ratio of the area of a circle and a square patch is $\pi R^2 \div$ (patch size), the fold within a circle of radius R reduces to (compare section 2.6)

$$\text{Fold} = \frac{(SD \times NC \times B^2) \times \pi R^2}{\text{Patch Size}}. \qquad (2.48)$$

R is equal to X_{mute}. If, however, X_{max} is measured along the diagonal of the patch, then no traces will be muted and X_{max} is indeed the maximum offset in the survey. In this case, all traces will be used in the stack (Figure 2.18b). This geometry makes fold calculations easier, and a more uniform fold distribution can be accomplished (see Sections 2.4, 2.5, and 2.6). Other aspects of patch design are discussed in Section 3.4. When X_{max} is measured in-line, the area of the patch is exactly twice the area of the patch with X_{max} measured diagonally.

The recorded X_{max} can be changed by placing source points off-center in the receiver patch. In this approach, the number of far offsets that exceed the "shorter far offsets" are reduced, which may not provide an adequate far-offset contribution.

In typical 3-D receiver patches, the mix of offsets is nonlinear, i.e., there are few near offsets and many far offsets (Figure 2.19). In fact, wide azimuth surveys have offset distributions that are linear in offset squared (see Section 3.3). Narrow azimuth surveys tend to have more linear offset distributions, with the offset distribution of a 2-D line being the end member. The dotted line in Figure 2.19 is the distribution for a

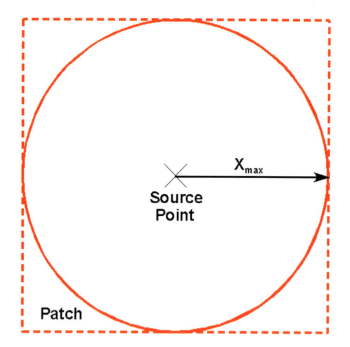

Fig. 2.18a. In-line X_{max}.

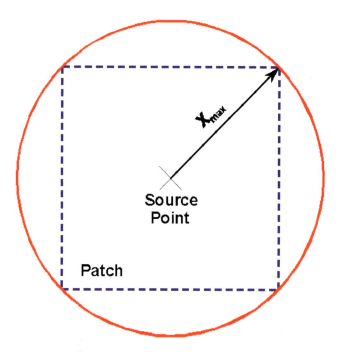

Fig. 2.18b. Diagonal X_{max}.

wide azimuth orthogonal design with eight receiver lines in the patch.

A good approximation (for a wide azimuth patch) is that the first 33% of the offset range includes 7% of

Table 2.7 Typical offset/fold distribution.

Offset	e.g.	Fold	e.g.
1/3	1000 m	10-20%	3-6
2/3	2000 m	50-70%	15-21
full	3000 m	100%	30

all offsets, the next 33% contributes 29% of the offsets, and the final 33% contains 64% of the offsets (N. Cooper, pers. comm.). Hence long offsets dominate the survey. Practical experience has shown that the amount of offsets representing each offset range in a wide azimuth survey result in fold variations as per Table 2.7. If X_{max} is set as the diagonal of the patch, the percentage of far offset traces (in the last third) is reduced to less than 50% of all traces in the recording patch (compare Figure 3.7a). The percentages in Table 2.7 are averages because the exact numbers depend on the aspect ratio and the number of receiver lines of the patch. Once the correct processing is performed (such as NMO, DMO, and migration), this nonlinear offset distribution should not pose a problem. However, array filtering and NMO stretch at long offsets could lead to lower overall frequencies.

To make good use of existing 2-D data to determine fold, S/N, X_{min}, and X_{max} is excellent advice. The off-

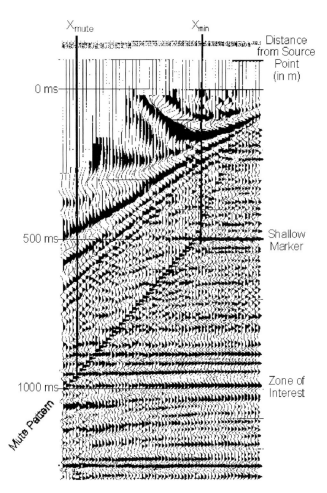

Fig. 2.20. NMO corrected common-shot gather.

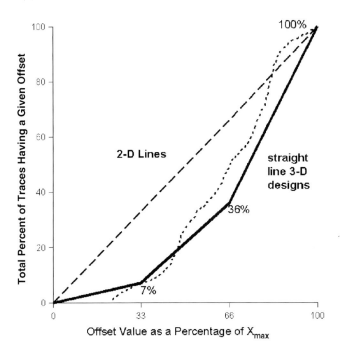

Fig. 2.19. Typical offset distributions (2-D versus 3-D).

set information (e.g., monitor records and NMO corrected common-shot gathers) of all available 2-D data should be examined thoroughly before deciding on the usable offset range for any 3-D seismic survey. Figure 2.20 shows an example where the maximum usable offset for imaging the zone of interest is 1400 m *(4600 ft)* as determined by the mute pattern. Traces were recorded out to 1800 m *(6000 ft)*, but these far-offset traces did not contribute to the stack at the targeted zone. Fold will therefore be lower than one might expect from fold calculations that are based on all offsets.

To determine X_{max}, one can prepare a geological model such as is shown in Figure 2.21a. Simple ray tracing reveals where reflected energy turns into refracted energy for each layer of interest. By examining a plot such as this, suitable values of X_{min} and X_{max}

34   *Planning and Design*

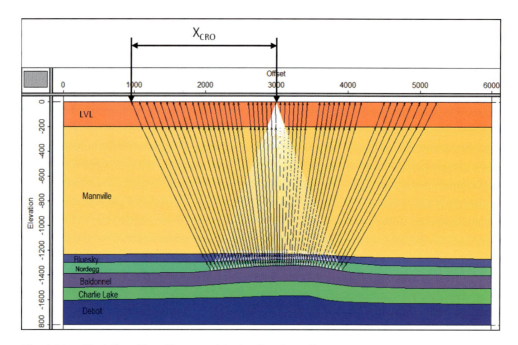

Fig. 2.21a. Modeling X_{max}. X_{CRO} = critical reflection offset.

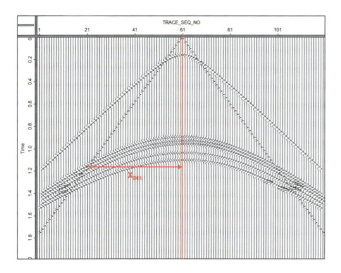

Fig. 2.21b. Synthetic shot for model above (50 m trace spacing).

can be determined for each event and hence for the entire model.

2.11.1 Target Depth

The largest offsets that should be recorded are generally the same magnitude as the target depth. Many models and survey data have shown this to be a good first approximation.

Rule of Thumb: X_{max} *should be approximately the same as the primary target depth, usually expressed as* X_{max} = *Target Depth.*

2.11.2 Direct Wave Interference

The direct wave, which travels from source to receiver by the shortest distance, will begin to interfere with primary reflection events at an offset X_{direct} and moveout time t_{NMO} (Figure 2.21b) where

$$X_{direct} = V_{LVL} \times (t_{NMO} + t_{mute}), \qquad (2.49)$$

and

$$t_{NMO} = \left(t^2 + \frac{X_{direct}^2}{V^2}\right)^{1/2}, \qquad (2.50)$$

where V_{LVL} is the velocity of material between source and receiver, V is the rms velocity to the primary event (target), and t_{mute} is a small mute zone (typically 200 ms) that is effectively the width of the direct arrival.

Substituting for t_{NMO},

$$X_{max} < X_{direct} = V_{LVL} \times \left[\left(t^2 + \frac{X_{direct}^2}{V^2}\right)^{1/2} + t_{mute}\right].$$

(2.51)

A computer program can solve these equations for the value of X_{direct}, given t, V, V_{LVL}, and t_{mute}. One should use existing 2-D data to determine possible near-surface effects. Direct wave interference also contributes to amplitude and phase artifacts in 3-D surveys, which may have deleterious effects on the acquired data. Note that good reflection data can exist in the region between the direct wave and the first breaks; therefore, this criterion should be used with caution.

2.11.3 Refracted Wave Interference (First Breaks)

The offset X_{ref} where the first break energy cuts across the primary reflection energy is given by,

$$X_{max} < X_{ref} = V_{ref} \times \left[\left(t^2 + \frac{X^2}{V^2}\right)^{1/2} + t_{mute}\right], \quad (2.52)$$

where V_{ref} is the velocity of a near-surface refracting layer. X_{ref} is always greater than X_{direct} (Figure 2.21b). Therefore, direct wave interference will always be more constraining than refracted wave interference.

2.11.4 Deep Horizon Critical Reflection Offset

Any trace at an offset greater than the critical reflection distance X_{CRO} for each layer will contain only refracted energy for that layer (Figure 2.21b). Adequate sampling of the layers with reflections is required. Thus a design constraint is:

$$X_{max} > X_{CRO}. \quad (2.53)$$

2.11.5 Maximum NMO Stretch To Be Allowed

NMO stretch is defined as the frequency distortion that occurs due to the normal move-out (NMO) correction. Software packages can model the percentage of NMO stretch versus time on each trace of a synthetic shot (see Yilmaz 1987, p. 160; equation 3.6). The patch design should include only offsets that create an NMO stretch smaller than the desired percentage (e.g., no more than 20%) at the target two-way time; that is,

$$NMO\ stretch = \frac{\Delta f}{f} = \frac{\Delta t_{NMO}}{t_0}, \quad (2.54)$$

where Δf is the change in frequency, f is frequency, Δt_{NMO} is the move-out time, and t_0 is the zero offset arrival time. The move-out depends upon the velocity as given by Yilmaz (1987):

$$\Delta t_{NMO} = t(x) - t(0)$$
$$= t(0)\left\{\left[1 + \left(\frac{X}{V_{NMO}\ t(0)}\right)^2\right]^{1/2} - 1\right\}. \quad (2.55)$$

2.11.6 Required Offset To Measure Deepest LVL (refractor)

At offset X_{deep}, the deepest low-velocity-layer satisfies the critical reflection criterion. To sample the LVL properly,

$$X_{max} > X_{deep}. \quad (2.56)$$

2.11.7 Required NMO Discrimination

One needs to record offsets X_{NMO} given by

$$X_{max} > X_{NMO} = V \times [(dt)^2 + 2 \times dt \times t_0]^{1/2}, \quad (2.57)$$

where dt = desired period (1.5 wavelengths), and t_0 = target time (deepest time).

2.11.8 Multiple Cancellation

One needs offsets X_{mult} given by

$$X_{max} > X_{mult} = V \times [(dt)^2 + 2 \times dt \times (t_{mult} \times 2)]^{1/2}, \quad (2.58)$$

where t_{mult} = multiple time (e.g., at $2 \times t_0$), V = multiple velocity, and dt = desired period at offset X_{mult}. At least three periods should be allowed because multiples are partially corrected by primary velocities at later times and will cancel only if more than one period remains.

2.11.9 Offsets Necessary for AVO/AVA

A range of offsets is needed where the angles of reflection from the target are sufficient to show the expected AVO effect (amplitude variation with offset due to the presence of gas or liquid). Narrow azimuth surveys have a better offset distribution for studying AVO effects. Amplitude variation with azimuth (AVA) work for the purpose of fracture detection requires sufficient offsets in many different azimuths.

2.11.10 Dip Measurements

The recorded offsets should be large enough to measure X_{max} as a function of dip. Reflections from any structural dipping layer are recorded at longer offsets on the downdip side than on the updip side. If one has a good knowledge of the expected dips, information can be used to make a geometrical correction to X_{max}.

Let's Design a 3-D—Part 2

In this example, these are the following known parameters (metric):

existing 2-D data of good quality have	30 fold
steepest dips	20°
shallow markers needed for isochroning have	500 m offsets
target depth	2000 m
target two-way time	1.5 s
basement depth	3000 m
V_{int} immediately above the target horizon	4200 m/s
f_{dom} at the target horizon	50 Hz
f_{max} at the target horizon	70 Hz
lateral target size	300 m

Orthogonal method is to be used.

Desired Fold:

$\frac{1}{2}$ up to full 2-D fold = 15 to 30

Bin Size:

a) for target size: B = 300 m ÷ 3 = 100 m
b) for alias frequency: $B = V_{int} \div (4 \times f_{max} \times \sin \theta)$
   = 4200 m/s ÷ (4 × 70 Hz × sin 20°) = 44 m
c) for lateral resolution: $B = V_{int} \div (N \times f_{dom})$ =
   4200 m/s ÷ (N × 50 Hz)
   = 21 to 42 m
bin size = 30 m × 30 m
RI = 60 m
SI = 60 m

Desired X_{min}:

RLI =
SLI =
$X_{min} = (RLI^2 + SLI^2)^{1/2}$ =

Desired X_{mute}:

number of channels in patch NC = fold × SLI × SI ÷ B^2 = _____

number of receiver lines _____
channels per line _____
cross-line dimension _____
in-line dimension _____
aspect ratio = cross-line dimension of the patch ÷ in-line dimension of the patch
$X_{max} = \frac{1}{2} \times $ [(in-line dimension of the patch)^2 + (cross-line dimension of the patch)^2]$^{1/2}$

Fold:

in-line fold = RLL ÷ (2 × SLI) =
cross-line fold = $\frac{1}{2}$ NRL =
total fold =

Let's Design a 3-D—Part 2 (Imperial)

In this example, these are the following known parameters:

existing 2-D data of good quality have	30 fold
steepest dips	20°
shallow markers needed for isochroning have	1500 ft offsets
target depth	6000 ft
target two-way time	1.5 s
basement depth	10 000 ft
V_{int} immediately above the target horizon	14 000 ft/s
f_{dom} at the target horizon	50 Hz
f_{max} at the target horizon	70 Hz
lateral target size	1000 ft

Orthogonal method is to be used.

Desired Fold:

$\frac{1}{2}$ up to full 2-D fold = 15 to 30

Bin Size:

a) for target size: B = 1000 ft ÷ 3 = 333 ft
b) for alias frequency: $B = V_{int} \div (4 \times f_{max} \times \sin \theta)$
   = 14 000 ft/s ÷ (4 × 70 Hz × sin 20°) = 146 ft
c) for lateral resolution: $B = V_{int} \div (N \times f_{dom})$
   = 14 000 ft/s ÷ (N × 50 Hz) = 70 to 140 ft
bin size = 110 ft × 110 ft
RI = 220 ft
SI = 220 ft

Desired X_{min}:

RLI =
SLI =
$X_{min} = (RLI^2 + SLI^2)^{1/2}$ =

Desired X_{mute}:

number of channels in patch $NC = fold \times SLI \times SI \div (B^2 \times U) = $ _____
number of receiver lines _____
channels per line _____
cross-line dimension _____
in-line dimension _____
aspect ratio = cross-line dimension of the patch ÷ in-line dimension of the patch
$X_{max} = \frac{1}{2} \times [(\text{in-line dimension of the patch})^2 + (\text{cross-line dimension of the patch})^2]^{1/2}$

Fold:

in-line fold $= RLL \div (2 \times SLI) = $
cross-line fold $= \frac{1}{2} NRL = $
total fold $=$

Chapter 2 Quiz

1. Which of the following factors affect in-line fold and cross-line fold, assuming no other changes in patch geometry?

			In-line	Cross-line
a.	X_{max}	Maximum offset	( )	( )
b.	RLI	Receiver line interval	( )	( ) within usable X_{mute}
c.	NRL	Number of receiver lines	( )	( )
d.	SLI	Source line interval	( )	( )
e.	B_s, B_r	bin size	( )	( )
f.	NC	Number of channels	( )	( )

3

Patches and Edge Management

3.1 OFFSET DISTRIBUTION

Figure 3.1 shows the relationships between offsets and azimuths. Each CMP bin usually contains midpoints from many source-receiver pairs; eight source-receiver pairs are shown contributing midpoints to this central bin. Each contributing trace in a bin has an offset (distance from source to receiver) and an azimuth (deviation from 0° north or compass angle) from source to receiver. For a successful 3-D survey, it is of paramount importance to consider both offset and azimuth.

Offset distribution in a stacking bin is most affected by fold. A lower fold gives poorer offset distribution, while increasing the fold improves the offset distribution. One must attempt to get an even offset distribution from near to far offsets to facilitate velocity calculations for normal moveout corrections and to obtain the best stacking response. A bad mix of offsets can cause aliasing of dipping signal, source noise, and perhaps even primary reflections.

Figure 3.2 shows one method of displaying the offset mix in each CMP bin. Each square is a CMP bin, and the number of sticks in each square equals the number of traces stacked in that bin. The vertical axis of each square shows the amount of the offset, and the horizontal axis indicates the position of the trace on an offset scale. In other words both vertical and horizontal scales are the offset value. A perfect triangular distribution of sticks would indicate the presence of all possible offsets. Two or more traces that have the same offset have the stick drawn in a different color to indicate redundancy.

As a designer, one should not be overly concerned about the offset and azimuth distribution in single bins. Migration and DMO move trace energy across many surrounding bins. What matters, therefore, is not the offset distribution of a single bin but rather the offset distribution in a "neighborhood" of bins. A good rule of thumb for the size of a neighborhood is the first Fresnel zone (see Section 3.4).

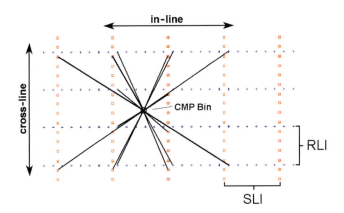

Fig. 3.1. Offsets and azimuths contributing to a CMP bin.

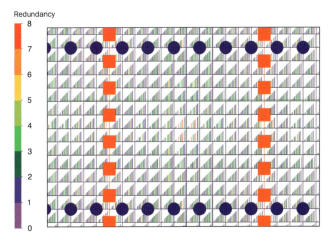

Fig. 3.2. Offset distribution—stick diagram.

40   Patches and Edge Management

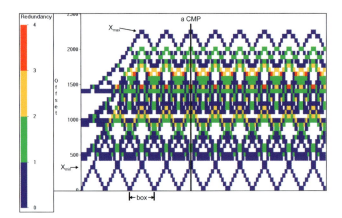

Fig. 3.3. Offset distribution in a row of bins.

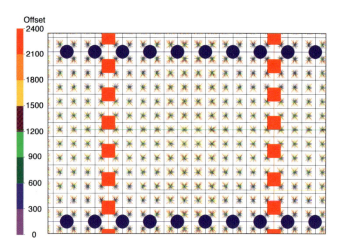

Fig. 3.4. Azimuth distribution—spider diagram.

A different method of displaying the mix of offsets in each bin is shown in Figure 3.3, which represents one line of bins. The horizontal scale is the CMP number and the vertical scale is offset. One CMP bin is represented with a vertical column. The vertical column is divided into small cells representing an offset range, usually chosen to be the group interval. The color bar of this figure indicates the number of repetitions of a particular offset in any given bin. Each cell is colored according to how many traces have an offset that lies in that bin.

In a display like this, the most uniform distribution of offsets in each CMP is indicated by a single color and by as many different offsets as possible in a set of neighboring bins (super bin). A fold of four or less for intervals equal to the group interval is acceptable. Therefore it is wise to select a color bar that highlights each fold value in a different color.

The distance from the base of one V to the base of the next is the width of a box in an orthogonal survey. X_{min} can be determined by zooming in on an area covered by a box and noting where the top of the V shape is located (e.g., at an offset of 200 m). This procedure should be repeated on several neighboring bin lines to obtain the largest value of X_{min}. The jump in the density of the fold in the bin-offset distribution of Figure 3.3 near 750-m offsets is caused by including two additional lines at the outside of the patch. For velocity analysis, all the possible offsets in this example are included by using all CMPs over an area equal to $\frac{1}{4}$ of a box (in a corner of a box, i.e., $\frac{1}{2}$ by $\frac{1}{2}$ of a box).

3.2   AZIMUTH DISTRIBUTION

Azimuth distribution in a stacking bin is most affected by fold, just as offset distribution is. If the aspect ratio of the patch is less than 0.5, one can expect a poor azimuth distribution. A bad mix of azimuths may lead to statics coupling problems and to an inability to detect azimuth-dependent variations that arise from dip and/or anisotropy. Increasing the aspect ratio to the range 0.6 to 1.0 solves such problems. A good azimuth distribution ensures that information from all angles surrounding the stacking bin is included in the stack.

Figure 3.4 shows a popular method (the spider plot) of displaying the azimuth of each trace that belongs to a midpoint bin. Each spider leg indicates the amount of offset (length and color of the leg) and points in the direction from source to receiver. The spider legs always start in the exact center of the bin and not necessarily at the midpoints; thus, this display does not show the midpoint scatter. The leg lengths are scaled to the largest offset in the entire survey, which is represented by a leg equal to half the bin dimension.

3.3   NARROW VERSUS WIDE AZIMUTH SURVEYS

The distinction between narrow and wide azimuth surveys is made on the basis of the aspect ratio of the recording patch. The aspect ratio is defined as the cross-line dimension of the patch divided by the in-line dimension. Recording patches with an aspect ratio less than 0.5 are considered narrow azimuth, while recording patches with an aspect ratio greater than 0.5 are wide azimuth.

Small-aspect-ratio patches (so-called narrow azimuth) lead to a more even distribution of offsets. However, these patches have, as the name indicates, a limited range of azimuths. Schematically, narrow azimuth surveys have a linear offset distribution with

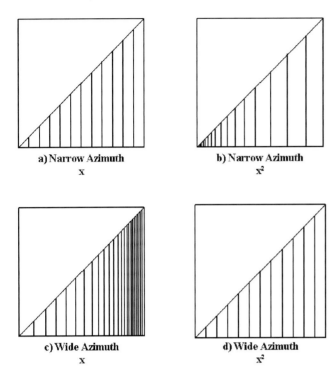

Fig. 3.5. Narrow- versus wide-azimuth templates and offset distributions. *X* = offset distance.

respect to offset similar to 2-D data (Figure 3.5a); however, when plotted against offset squared, the offset distribution shows bunching at the near offsets (Figure 3.5b). Narrow azimuth patches are better for AVO and DMO purposes and when significant lateral velocity variations are present (Lansley, 1994).

Wide-azimuth surveys (i.e., patches that are closer to a square) have a nonlinear offset distribution with respect to x, with a heavy weighting of the far offsets (Figure 3.5c). However, when plotted against offset squared, the distribution is nearly linear (Figure 3.5d). Wide-azimuth surveys are better for velocity analysis, multiple attenuation, static solutions, and a more uniform directional sampling of the subsurface. These diagrams are schematic, and variations in offset distributions may occur in real data.

Comparisons have been made to illustrate differences between narrow and wide azimuth acquisition. Figures 3.6, 3.7, and 3.8 compare a wide-azimuth acquisition consisting of a patch of 12 lines with 60 stations per line in the left column (a, c, e) to a narrow-azimuth acquisition comprised of a patch of 6 lines with 120 stations per line in the right column (b, d, f). The respective aspect ratios are 0.80 and 0.20 based on station spacings of 60 m and receiver line spacings of 240 m.

Since neither the source point density nor the number of receivers in the patch have been changed, the nominal fold for both acquisition strategies (Figures 3.6a, b) is 30 (Figures 3.6c, 3.6d). What does change, however, is the configuration of the fold taper, which is much less in the cross-line direction for the narrow azimuth patch (Figure 3.6d). The fold at an offset limit of 1500 m is significantly lower for the narrow azimuth patch (the fold scale is constant in Figures 3.6e, 3.6f). The comparison of the offset distribution shows that the wide-azimuth patch has traces closer to the source points than the narrow-azimuth patch (Figures 3.7a, 3.7b), assuming that the same number of receivers are utilized in the patch. An average trace count of the narrow patch was copied onto the wide-patch display (Figure 3.7a) with the corrected scaling. The stick diagrams (Figures 3.7c, 3.7d) and the offset variation within a box (Figures 3.7e, 3.7f) indicate that the offset distribution is better for the wide patch because of the nonlinearity in the source–receiver spacings that results from the azimuthal distribution of the receivers. An azimuth-dependent trace count shows that there is a more even distribution of source–receiver pairs for the wide patch (Figures 3.8a, 3.8b). The azimuth distribution is far more varied for the wide patch than for the narrow patch (Figures 3.8c, 3.8d). The rose diagram in Figures 3.8e and 3.8f uses color to indicate the multiplicity of the occurrence of a particular source–receiver pair in offset and azimuth distribution (for the entire survey) and shows the focused nature of the narrow-azimuth patch.

3.4 85% RULE

Three-dimensional crews often record with more channels on the ground than are necessary. This practice extends the recorded X_{max} beyond the required X_{mute}. Equipment availability should be taken into consideration when the patch size is determined.

If a wide-azimuth survey is desired, one must decide on the best aspect ratio for the patch. For the moment, this discussion is restricted to square patches with an aspect ratio of 1.0, meaning that the in-line dimension equals the cross-line dimension. Consider a circle of unit area (i.e., = 1.0) with a radius of X_{mute} representing the mute zone (large red circle in Figure 3.9). If the patch lies entirely outside this circle, then 27% of the channels in the patch are being used to record data that will most likely be muted out. While these channels may have some value for longer wavelength refraction analysis, using that many extra channels may be expensive. The recorded X_{max} is a factor of $\sqrt{2}$ larger than the required X_{mute}.

42   *Patches and Edge Management*

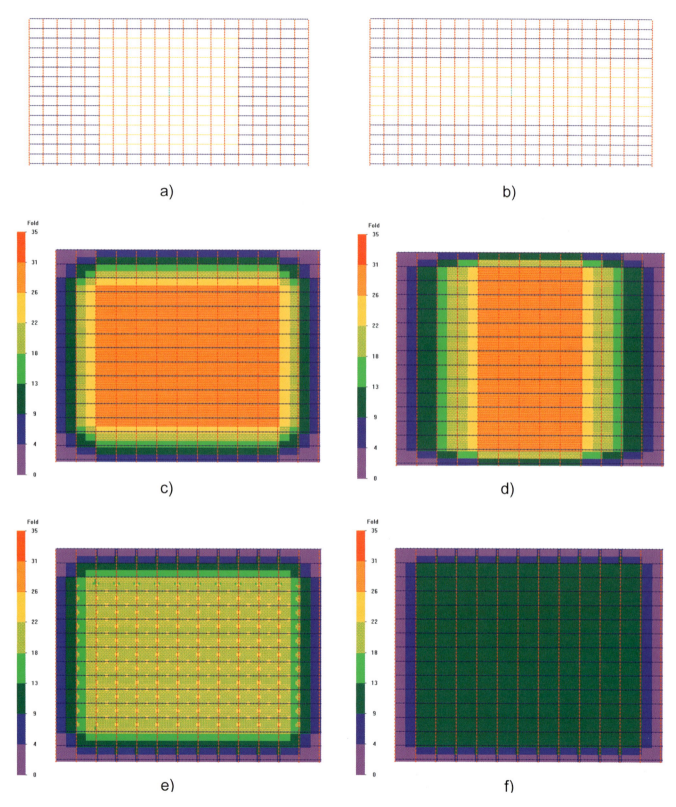

Fig. 3.6. Wide versus narrow patch—template and fold; a. wide-azimuth template, b. narrow-azimuth template, c. wide-azimuth fold distribution at full offsets, d. narrow-azimuth fold distribution at full offsets, e. wide-azimuth fold distribution at 1500 m offsets, and f. narrow-azimuth fold distribution at 1500 m offsets.

3.3 *Narrow versus Wide Azimuth Surveys* 43

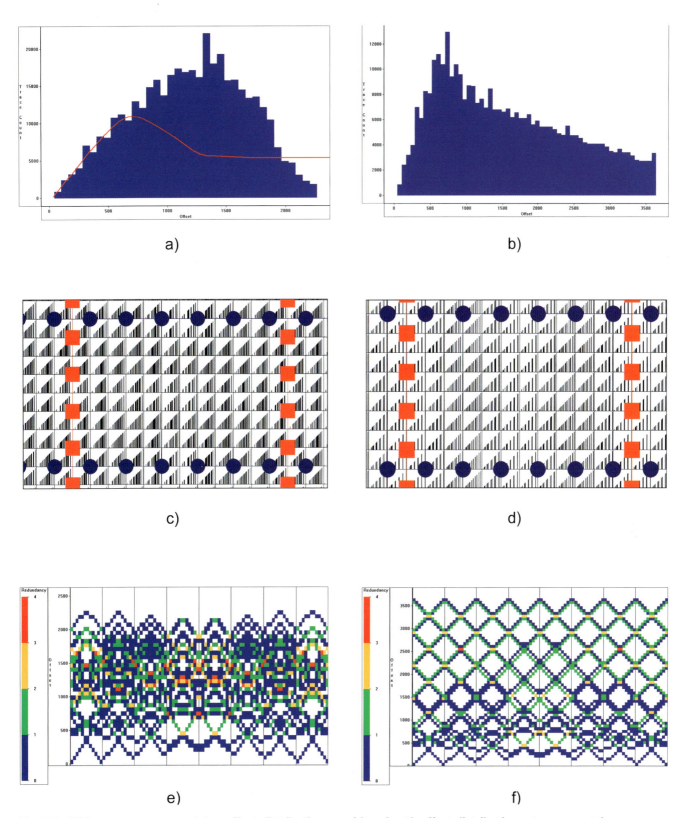

Fig. 3.7. Wide versus narrow patch—offset distribution; a. wide-azimuth offset distribution—trace count, b. narrow-azimuth offset distribution—trace count, c. wide-azimuth offset distribution—stick diagram, d. narrow-azimuth offset distribution—stick diagram, e. wide-azimuth offset distribution—offset variation within a box, and f. narrow-azimuth offset distribution—offset variation within a box.

44   *Patches and Edge Management*

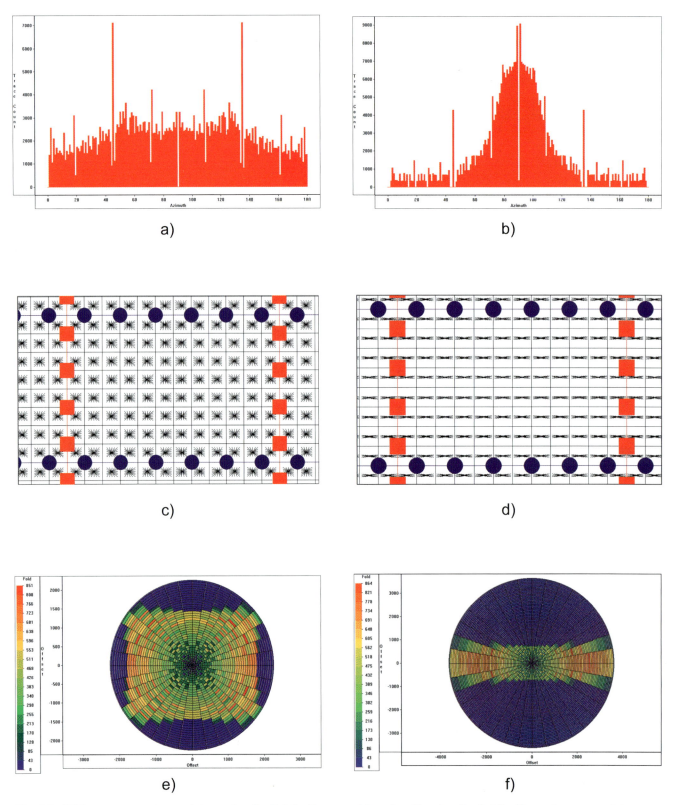

Fig. 3.8. Wide versus narrow patch—azimuth distribution; a. wide-azimuth azimuth distribution—trace count, b. narrow-azimuth azimuth distribution—trace count, c. wide-azimuth azimuth distribution—spider diagram, d. narrow-azimuth azimuth distribution—spider diagram, e. wide-azimuth azimuth distribution—rose diagram, and f. narrow-azimuth azimuth distribution—rose diagram.

On the other hand, one can reduce the patch to lie entirely within the mute zone, as shown by the small blue square. X_{max} is now measured along the diagonal of the patch, but this patch covers only 64% of the area of the design objective, i.e., the large red circle. This is the other extreme of inefficiency; there are only a few traces that lie at offsets close to the ideal mute distance. The recorded X_{max} equals the required X_{mute} in this case. Some companies select a patch size that equals the large red square in order to record offsets of X_{mute} in all directions. However, the large red square is twice the area of the small blue square, and there is a significant difference in cost and effort between the two patches.

The 85% Rule is a compromise and determines the aspect ratio of the patch relative to the desired X_{mute}. The 85% Rule is a simple way to optimize the area of usable traces recorded and the number of channels needed. The rule works as follows (Figure 3.10):

1. Determine the desirable X_{mute}.
2. Choose the in-line offset X_r to be $0.85 \times X_{mute}$.
3. Choose the cross-line offset X_s to be $0.85 \times X_r = 0.72 \times X_{mute}$.

For a real example with X_{mute} = 2000 m,   (6600 ft)
half in-line dimension
$\quad X_r \quad = 85\% \times X_{mute} = 1700$ m,   (5610 ft)
half cross-line dimension
$\quad X_s \quad = 85\% \times X_r \quad = 1445$ m,   (4730 ft)
aspect ratio
$\quad X_s \div X_r \quad = 85\%$.

The usable area of the patch relative to the circle of X_{mute} increases from 64% to 78%. Only a small part of the patch is outside the theoretical maximum offset X_{mute} (e.g., for a 6-line patch the traces at offsets greater than X_{mute} are less than 2.5%). Additional receiver lines farther out than indicated by this patch are mostly outside the usable offset X_{mute}. Therefore, the longer dimension of the patch is preferred to be in the in-line direction. The dimensions may need some slight adjustment to be suitable for other considerations in the design of the 3-D. The recorded X_{max} is 1.13 times the required X_{mute}; most likely the mute affects only the far extremes of the two farthest receiver lines.

Referring back to Figure 3.9, the relationships between the different areas are shown graphically in Figure 3.11. The red circle of radius X_{mute} has unit area (100% at 100% X_{mute}) and is shown as the solid line. The inner blue circle with a radius of 0.71 X_{mute} contains 50% of the area of the larger circle. If the patch is entirely within X_{mute} (inner blue square in Figure 3.9), then the curve describing its area deviates from the $y = x^2$ form at 0.71 X_{mute}. The area of this smaller blue patch is 64% of the unit area (red circle).

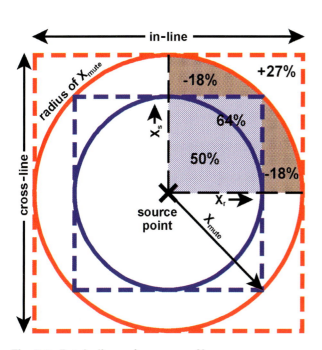

Fig. 3.9. Patch dimension versus X_{mute}.

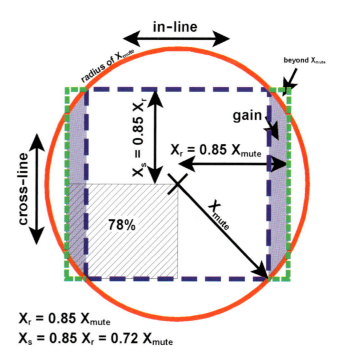

Fig. 3.10. Ideal patch, using the 85% rule.

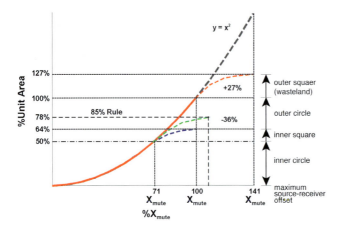

Fig. 3.11. Percentage of area covered for various choices of X_{mute}.

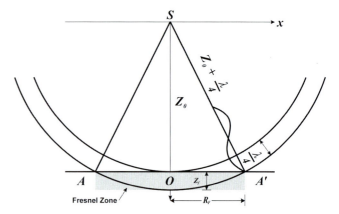

Fig. 3.12. Fresnel zone (after Yilmaz, 1987).

On the other hand, if the patch is entirely outside the red circle of X_{mute} (larger red square in Figure 3.9), then the curve describing its area deviates from the $y = x^2$ trend at 100% X_{mute}. The area of this larger patch is 127% of the unit area, which is twice the area of the smaller patch, with its maximum offsets being 141% of the mute offset. Therefore if one lays out receivers in a square patch that has X_{max} measured in the in-line direction, then twice the number of receivers are needed in comparison to a patch that has X_{max} measured along the diagonal.

The patch using the 85% rule covers 78% of the unit area described by the red circle of X_{mute}. The green patch covers an area that is 22% larger than the smaller blue square (compare Figure 3.10). The patch using the 85% rule is an excellent compromise for the patch design with the recorded maximum offsets being 113% of the required mute zone. The design provides a sufficient number of large offsets, and only a few traces may have to be deleted in the stack.

3.5 FRESNEL ZONE

The Fresnel zone is that area of a reflector from which reflected energy can reach a detector and not be more than one-half wavelength out of phase with any other energy reflected from within that area (Sheriff, 1991). In Figure 3.12, energy traveling from a source at S to a subsurface point 0 arrives at the surface at time $t_0 = 2 \times Z_0 \div V_{ave}$. Reflected energy from point A or A' will reach the receiver at point 0 at time $t_1 = 2[(Z_0 \div V_{ave}) + (\lambda \div 4V_{ave})]$. All energy arriving within the time interval $(t_1 - t_0)$ interferes constructively.

The reflecting disk AA' is called the first Fresnel zone (Sheriff, 1991). Two reflecting points that fall within this zone are generally considered indistinguishable when observed at the earth's surface. The radius of the first Fresnel zone R_F for vertical incidence can be approximated as

$$R_F = \left(\left(Z_0 + \frac{\lambda}{4}\right)^2 - Z_0^2\right)^{1/2} = \left(\frac{Z_0 \times \lambda}{2} + \frac{\lambda^2}{16}\right)^{1/2}, \tag{3.1}$$

which can be further approximated by,

$$R_F \cong \left(\frac{Z_0 \times \lambda}{2}\right)^{1/2} = \left(\frac{t_0 \times V_{ave}}{2} \times \frac{V_{ave}}{2f_{dom}}\right)^{1/2} \tag{3.2}$$

or

$$R_F \cong \tfrac{1}{2} V_{ave} \left(\frac{t_0}{f_{dom}}\right)^{1/2},$$

where f_{dom} is the dominant frequency being considered. At lower frequencies, the Fresnel zone becomes larger, while for higher frequencies the Fresnel zone is smaller. This derivation follows the approach of Yilmaz (1987). For nonzero offset, structure and dip affect the shape of the Fresnel zone. Lindsey (1989) points out that the Fresnel zone becomes somewhat smaller for positive structures such as reefs or anticlines, and somewhat larger for negative structures such as synclines. Generally, this is a second-order effect except when structures are of similar scale to the depth.

The vertical thickness of the Fresnel zone Z_F is

$$Z_F = \frac{\lambda}{4} = \frac{V_{ave}}{4f_{dom}}. \tag{3.4}$$

Expressing this in time, the two-way time thickness of the Fresnel zone T_F becomes

$$T_F = \frac{\frac{2 \times \lambda}{4}}{V_{\text{ave}}} = \frac{\lambda}{2V_{\text{ave}}}, \quad (3.5)$$

where λ is the wavelength of the dominant frequency.

The diameter of the Fresnel zone determines the lateral resolution before migration. In the context of diffractions, lateral resolution can be defined as the ability to distinguish two adjacent diffractions. Because migration is a process that collapses diffractions, it is reasonable to think that migration increases spatial resolution. In effect, migration moves the plane of observation downward to the reflection points. Therefore, the Fresnel zone becomes smaller (Yilmaz, 1987). Migration of 3-D data tends to collapse the Fresnel zone diameter to approximately one-half the dominant wavelength, while migration of 2-D data accomplishes this only in the direction of the seismic line (Lindsey, 1989).

Migration introduces unwanted artifacts near the edges of the 3-D survey because part of the Fresnel zone lies outside the survey. These artifacts are of particular importance when undershooting corners of lease blocks.

The above discussion considers only monochromatic signals. Knapp (1991) proposed a generalization of the Fresnel zone definition for broadband signals that guarantees the correct wavelet and amplitude of the reflected signal. Brühl et al. (1996) determined that by comparing monochromatic, narrow band, and broadband signals that the boundary of the (first) Fresnel zone corresponds to the position of maximum energy build-up. Brühl et al. (1996) extended Knapp's concept to a "zone of influence," because the size of the area that guarantees the correct wavelet depends on the length of the wavelet. By definition, the zone of influence is the area on the reflector for which the difference between the reflection traveltimes and the diffraction traveltimes is less than the length of the wavelet (rather than one half of the wavelength). This definition separates the reflected wavelet from the input wavelet. Any restriction of the reflector area to a radius smaller than that of the zone of influence would result in a change of the reflected wavelet with respect to the input wavelet.

3.6 DIFFRACTIONS

Diffractions occur where sharp reflector boundaries and discontinuities such as faults are present. Diffractions extend far beyond and especially perpendicular to faults. The migration apron needed to span these diffractions adds to the surface coverage needed to image the subsurface properly. Therefore, calculations of migration apron need to be made prior to starting any 3-D design.

When source energy reaches a subsurface point discontinuity (point diffractor), the reflected energy can be ray traced upward to the surface at all angles. Wherever a receiver is positioned on the surface, it records a reflection corresponding to the time it takes for the energy to travel from the source to the diffractor plus the time it takes to travel upward. The collection of all these various reflections along the receiver line creates a diffraction curve that has interesting properties.

The area at the top of the diffraction curve that has a thickness equal to one-quarter wavelength of the dominant frequency (or a two-way interval time equal to one-half cycle of the dominant frequency) is commonly called the first Fresnel zone. Other points on the diffraction curve correspond to different rays traced from the diffractor to the surface at varying angles. Therefore, that part of the curve near its apex corresponds to low scattering angles, and the parts of the curve at longer two-way times correspond to greater angles.

After migration, the diffraction curve (actually a bell shape in 3-D) is collapsed, not quite to a spike as described earlier, but to the highest frequency wavelet in time and space permitted by the signal bandwidth. If only a portion of the diffraction curve can be used, the collapse will not be quite complete, and a migrated version of the event will contain only a fraction of the correct energy.

Using only the first Fresnel zone portion of a diffraction curve gives approximately 70% of the energy of the fully migrated result (the Fresnel zone corresponds approximately to a 15° scattering angle). Using both sides of the diffraction out to a scattering angle of 30° gives approximately 95% of the fully migrated result (Figure 3.13). The exact quantity (e.g., 95%, 96%, ...) depends on the velocities and depths of the diffractors. The recording of a complete diffraction curve is not necessary to get a useful migrated result.

3.7 MIGRATION APRON

Migration is necessary to place dipping horizons and faults at their proper subsurface positions. When designing the boundaries of a survey, the full-fold area covered must be increased to allow for the migration apron. The amount of increased area is not

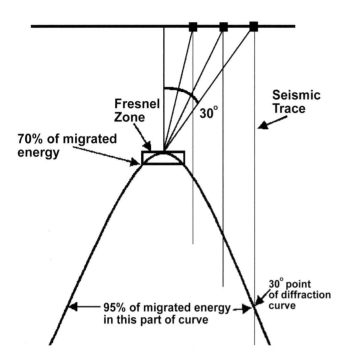

Fig. 3.13. Anatomy of a diffraction curve.

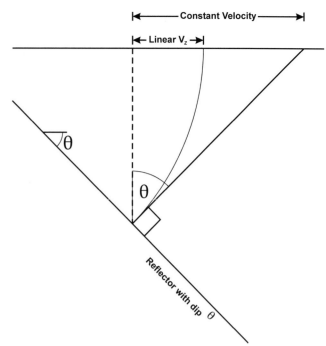

Fig. 3.14. Width of migration apron for media having a constant velocity field versus a linearly varying velocity (after Bee et al., 1994).

necessarily the same in the strike and the dip direction. In a constant-velocity medium, dipping horizons require a migration apron given by the following formula:

$$MA = Z \tan \theta, \qquad (3.6)$$

where MA = migration apron, Z = depth, and θ = dip angle (true reflector dip).

Assuming straight raypath geometry, if one wants to record reflections from a horizon at a depth of 2000 m (*6500 ft*) that dips 20°, then 728 m (*2366 ft*) need to be added to the 3-D grid size as the migration apron. However, to capture the 30° ray off the outermost point of a diffraction, the migration apron must be at least $Z \times \tan 30° = 0.58 \times Z$ as per Figure 3.13 (in the above example, 0.58 × 2000 m = 1155 m or *0.58 × 6500 ft = 3753 ft*). This condition defines the desired migration apron unless dips exceed 30°. Table 3.1 has constant migration apron values in the columns for 10°, 20°, and 30° because the diffraction criterion is the constraining factor. In practice, cost considerations often force a compromise on this desired apron. Curved raypaths assumptions reduce the required migration apron (Figure 3.14; Bee et al., 1994), especially if large dips are present. The depth-varying velocity field should be taken into account when calculating the size of the migration apron. Ray tracing

from an exploding reflector at depth can help determine the required migration apron in a velocity layered medium.

Rule of Thumb: Migration apron is normally chosen as the larger of:

1) *The lateral migration movement of each dip in the expected geology, or*
2) *The distance required to capture diffraction energy coming upwards at a scattering angle of 30°, or*
3) *The radius of the first Fresnel zone.*

Displays such as Figure 3.15 help determine vertical resolution (one-quarter wavelength of dominant frequency) and lateral resolution before migration (Fresnel zone diameter versus bin size). Other displays allow the selection of the bin size required to image a desired high frequency on certain dips in the model. The survey size should be calculated based on the area to be imaged plus the migration apron.

3.8 EDGE MANAGEMENT

Edge management refers to the aspect of 3-D design that specifies the width of the migration apron and the fold taper. The fold taper is the area at the edge of

Table 3.1. Migration apron calculations for a variety of target depths and structural dips.

a) Metric

Diffractor depth (m)	10°	20°	30°	40°	50°	60°	70°	80°	90°
500	289	289	289	420	596	866	1374	2836	∞
1000	577	577	577	839	1192	1732	2747	5671	∞
1500	866	866	866	1259	1788	2598	4121	8507	∞
2000	1155	1155	1155	1678	2384	3464	5495	11343	∞
2500	1443	1443	1443	2098	2979	4330	6869	14178	∞
3000	1732	1732	1732	2517	3575	5196	8242	17014	∞
3500	2021	2021	2021	2937	4171	6062	9616	19849	∞
4000	2309	2309	2309	3356	4767	6928	10990	22685	∞
4500	2598	2598	2598	3776	5363	7794	12364	25521	∞
5000	2887	2887	2887	4195	5959	8660	13737	28356	∞

← Diffraction →   ← Migration of dipping events →

b) Imperial

Diffractor depth (ft)	10°	20°	30°	40°	50°	60°	70°	80°	90°
1000	577	577	577	839	1192	1732	2747	5671	∞
2000	1155	1155	1155	1678	2384	3464	5495	11343	∞
3000	1732	1732	1732	2517	3575	5196	8242	17014	∞
4000	2309	2309	2309	3356	4767	6928	10990	22685	∞
5000	2887	2887	2887	4195	5959	8660	13737	28356	∞
6000	3464	3464	3464	5035	7151	10392	16485	34028	∞
7000	4041	4041	4041	5874	8342	12124	19232	39699	∞
8000	4619	4619	4619	6713	9534	13856	21980	45370	∞
9000	5196	5196	5196	7552	10726	15588	24727	51042	∞
10000	5774	5774	5774	8391	11918	17321	27475	56713	∞
11000	6351	6351	6351	9230	13109	19053	30222	62384	∞
12000	6928	6928	6928	10069	14301	20785	32970	68055	∞
13000	7506	7506	7506	10908	15493	22517	35717	73727	∞
14000	8083	8083	8083	11747	16685	24249	38465	79398	∞
15000	8660	8660	8660	12586	17876	25981	41212	85069	∞

← Diffraction →   ← Migration of dipping events →

the survey where full-fold (or nearly full-fold) at depth has not been reached before migration. Often a small relaxation in the definition of sufficient fold near the edges can significantly reduce the total size of the survey.

The fold taper can easily add 30% to the total 3-D area, even on large surveys. On small surveys, this percentage increases disproportionately and makes small 3-D surveys, which might be intended to cover as little as 2.5 km² *(1 mi²)*, extremely expensive on a cost-per-area basis. The design of the patch (and therefore fold taper) is often different in the receiver line direction than in the source line direction because dips may change depending on the azimuth, and then the migration apron changes with dip azimuth. Calculations of the fold taper distances were introduced in Chapter 2. Figure 3.16 shows the usable 3-D program size (net of the fold taper) as a percentage of the total program size for a target depth of 2000 m *(6500 ft)* and fold taper widths of 200 to 1600 m *(660 to 5280 ft)*.

Generally, fold builds up faster in the cross-line direction than in the in-line direction. This is especially true for narrow azimuth designs because the fold taper in the cross-line direction is directly proportional to the cross-line fold. Careful distribution of receivers can build up fold faster for certain geometries. The orientation of the patch with respect to the survey outline can affect the cost of a survey significantly.

50   *Patches and Edge Management*

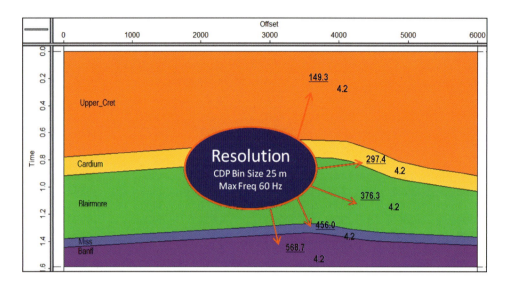

Fig. 3-15. Modeling for vertical and horizontal resolution.

Any 3-D survey should be considered as consisting of three zones (Figure 3.17). The first (innermost) zone is the domain of the interpreter. All traces lying in this zone should be considered full-fold and fully migrated. This is the image area the interpreter should limit his examinations to and use as the basis for geological interpretation. The second (middle) zone is a corridor around the innermost (image) zone. Theoretically, the width of this corridor is equal to the migration apron. In this corridor, the seismic processor assembles full-fold stacked traces. Migration moves most of the energy of these traces into the edge of the innermost (image) zone. The third (outermost) zone is a corridor around the middle zone. The width of this zone is the fold taper. In this corridor, the acquisition

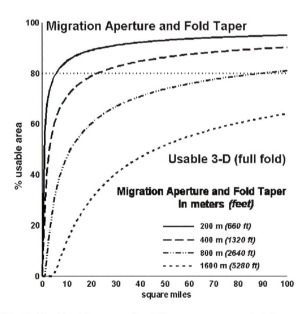

Fig. 3.16. Usable area of a 3-D survey versus total recorded area.

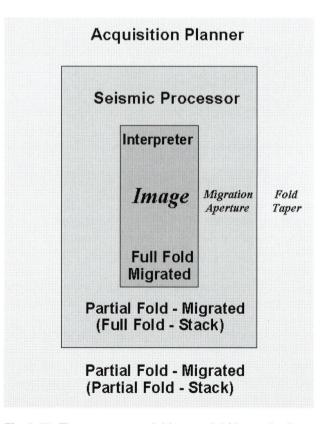

Fig. 3.17. Three-zone acquisition model (theoretical).

3.9 *Ray-Trace Modeling* 51

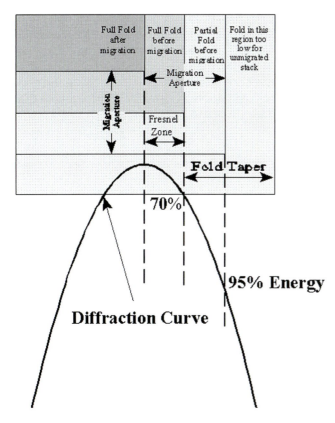

Fig. 3.18. Practical edge management.

planner places sources and receivers to ensure full-fold at the start of the middle (full-fold) zone. The term edge management implies the proper design of these three zones. Compromises can and almost always will be made.

It is not necessary that traces in the middle zone have near offsets. Farther traces contain deep data because of NMO muting, but deep data migrate farther laterally. If the bins near the outside of the second zone contain only far traces, and bins closest to the innermost zone contain more near traces (hence shallow data), a good migrated result on the edge of the innermost (image) zone should result at all depths.

It may also be possible to relax the 30° requirement for very deep data. This concession reduces the size of the migration apron (Figure 3.18). Traces that are not quite full-fold may be acceptable for bins near the outside edge of the middle zone. These traces correspond to the near 30° portion of the diffraction and do not contribute much energy at the edge of the innermost (image) zone.

Rule of Thumb: In no case should the width of the total migration apron around the image area be less than the Fresnel zone radius plus the full-fold taper size.

This design rule means there are full-fold traces from the edge of the image area outward to one Fresnel zone radius and lesser fold from there to the edge of the survey. Most often, the migration apron and the fold taper blend together, and cost considerations influence the final choices for a particular design. Experiments with models help identify these parameters in any specific project.

3.9 RAY-TRACE MODELING

So far, the term CMPs or common-midpoints has been used. However, the energy returning from a reflector does not necessarily originate from that point on the reflector that is halfway between source and receiver. Migration corrects traces from their CMP position to the CDP (common-depth-point) position. It is important to know how well each piece of the target is illuminated, and not only what fold is achieved after migration. This imaging concept involves CDP fold. Three-dimensional ray tracing is essential for true CDP analysis.

Ray-trace modeling (Figure 3.19) is useful if the underlying geology is more complicated than the flat-layer model, and is often used for 3-D design. Examples of situations where ray tracing should be done include salt domes, faults, steeply dipping layers, and lateral velocity discontinuities. Such modeling may lead the designer to a different strategy for the surface layout than do the flat-layer assumptions. Source and receiver spacing may be reduced or increased in certain areas of the seismic survey to assure coverage in structurally complex areas (Neff and Rigdon, 1994). Sophisticated computer programs are available to evaluate illumination fold distributions versus nominal fold in structured

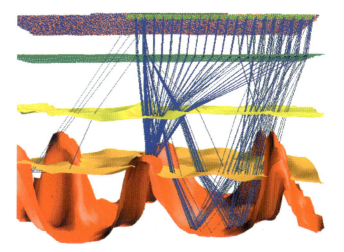

Fig. 3.19. Rays belonging to one source point after 3-D ray tracing of a salt layer (Cain et al., 1998).

52 Patches and Edge Management

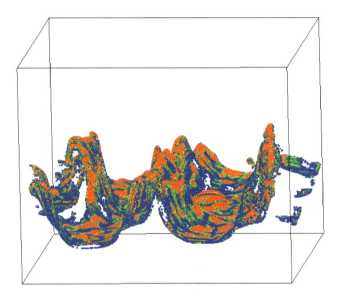

Fig. 3.20. 3-D illumination fold map for salt layer (Cain et al., 1998).

environments (Cain et al., 1998). The fold can be draped over the structural elements of Figure 3.19 and the illumination is evaluated for the target horizons (Figure 3.20). Some salt flanks have not been illuminated at all, which hinders the interpretation severely. The impact of varying fold and offset and azimuth distributions can be rigorously evaluated with these programs.

3.10 RECORD LENGTH

The total record length needs to include several arrival times. First, the vertical two-way time to the main target can easily be obtained from existing 2-D data. Second, the record length should be chosen so that any diffraction patterns from the deepest event of interest (which may include basement) are properly recorded suitable for migration.

Margrave (1997) summarized these ideas for constant-velocity situations (Figure 3.21) and showed that the record length has to be at least:

$$t = \frac{2Z}{V \cos \theta}, \quad (3.7)$$

where θ is the scattering angle between the vertical axis and the farthest receiver to be recorded.

Third, the moveout times for reflectors and multiples have to be taken into account at the far offsets. Finally, static shifts may require up to 100 ms more and instrumentation requirements add an additional 100–200 ms. Therefore the total required record length is significantly more than simply the two-way time to the main target.

For example, assume that existing 2-D data show that the target horizon is at 1.5 s. Furthermore, assume that it is desirable to image basement, which is at 2.5 s. Diffraction tails are on the order of 500 ms, moveout is assumed to add 300 ms, dip requirements may add 400 ms, static shifts may require up to 100 ms, and instrumentation requirements are 100 to 200 ms. Although there is some overlap among these requirements, one would probably choose a recording time of 3.5 to 4.0 s.

It is always easy to record longer trace data because tape is usually cheap in comparison to other recording costs. The only concern might be with telemetry systems, where a longer record length may slow down the overall acquisition effort because of the necessity

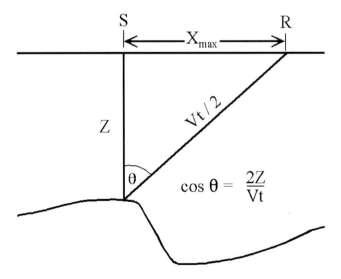

Fig. 3.21. Record length t as a function of scattering angle θ (after Margrave, 1997).

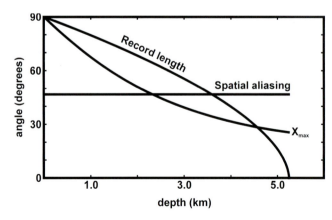

Fig. 3.22a. Constant velocity scattering angle chart showing: aperture limit, record length limit, and spatial aliasing limit for a case when X_{max} = 2500 m, t = 3.0 s, V = 3500 m/s, B = 20 m, and f = 60 Hz (after Margrave, 1997).

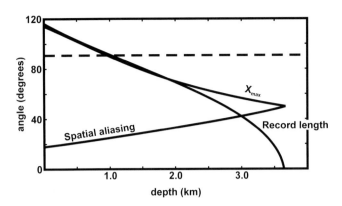

Fig. 3.22b. Constant gradient scattering angle chart showing: aperture limit, record length limit, and spatial aliasing limit for a case when X_{max} = 2500 m, t = 3.0 s, V = 1500 m/s + kZ^c with k = 8 Hz and c = 0.6; B = 20 m, and f = 60 Hz (after Margrave, 1997).

to transmit the information from each station sequentially after the shot has been taken. This limitation has been reduced with recent product enhancements.

Margrave (1997) has described how the record length, the far-offset limit (as defined by X_{max}), and spatial aliasing limit the maximum scattering angle that can be recorded for the constant velocity case and straight raypaths (Figure 3.22a), as well as the depth varying velocity case V_z and curved raypaths (Figure 3.22b). Such displays can be created for other velocity fields, recording times, maximum offsets, bin sizes, and maximum frequencies.

Let's Design a 3-D – Part 3 (metric)

In this example, these are the following known parameters:

existing 2-D data of good quality have	30 fold
steepest dips	20°
shallow markers needed for isochroning have	500 m offsets
target depth	2000 m
target two-way time	1.5 s
basement depth	3000 m
V_{int} immediately above the target horizon	4200 m/s
f_{dom} at the target horizon	50 Hz
f_{max} at the target horizon	70 Hz
lateral target size	300 m

Orthogonal method is to be used.

Desired Fold:

$\frac{1}{2}$ up to full 2-D fold = 15–30

Bin size:

a) for target size: B = 300 m ÷ 3 = 100 m
b) for alias frequency: $B = V_{int} ÷ (4 \times f_{max} \times \sin\theta)$
= 4200 m/s ÷ (4 × 70 Hz × sin 20°)
= 44 m
c) For lateral resolution: $B = V_{int} ÷ (N \times f_{dom})$ = 4200 m/s ÷ (N × 50 Hz) = 21 to 42 m
bin size = 30 m × 30 m
RI = 60 m
SI = 60 m

Desired X_{min}: 500 m

RLI = 360 m
SLI = 360 m
$X_{min} = (360^2 + 360^2)^{1/2}$ m = 509 m
(467 m with one bin offset at line intersections)

Desired X_{mute}: 2000 m

patch 8 × 60 or 2880 m × 3600 m
number of channels = 480
aspect ratio = 2880 ÷ 3600 = 0.80
$X_{max} = \frac{1}{2} \times (2880^2 + 3600^2)^{1/2}$ m = 2305 m

Fold:

in-line fold = 60 m × 60 m ÷ (2 × 360 m) = 5
cross-line fold = 8 ÷ 2 = 4
total fold = 20

Migration Apron:

radius of Fresnel Zone =
$\frac{1}{2} \times V_{ave} \times$ (target two-way time ÷ f_{dom})$^{1/2}$ =
diffraction energy = 0.58 × target depth =
migration apron = target depth × tan (dip) =
in-line fold taper =
[(in-line fold ÷ 2) − 0.5] × SLI =
cross-line fold taper =
[(cross-line fold ÷ 2) − 0.5] × RLI =
(FT + FZ) < total migration apron < (FT + MA)
TMA range =

Let's Design a 3-D—Part 3 (imperial)

In this example, these are the following known parameters:

existing 2-D data of good quality have	30 fold
steepest dips	20°
shallow markers needed for isochroning have	1500 ft offsets
target depth	6000 ft
target two-way time	1.5 s

basement depth	10 000 ft
V_{int} immediately above the target horizon	14 000 ft/s
f_{dom} at the target horizon	50 Hz
f_{max} at the target horizon	70 Hz
lateral target size	1000 ft

Orthogonal method is to be used.

Desired Fold:

$$\tfrac{1}{2} \text{ up to full 2-D fold} = 15-30$$

Bin size:

a) for target size: B = 1000 ft ÷ 3 = 333 ft
b) for alias frequency: B = V_{int} ÷ (4 × f_{max} × sin θ)
   = 14 000 ft/s ÷ (4 × 70 Hz × sin 20°) = 146 ft
c) for lateral resolution: B = V_{int} ÷ (N × f_{dom}) =
   14 000 ft/s ÷ (N × 50 Hz)
   = 70 to 140 ft
bin size = 110 ft × 110 ft
RI = 220 ft
SI = 220 ft

Desired X_{min}: ~1500 ft

RLI = 880 ft
SLI = 1320 ft
X_{min} = $(880^2 + 1320^2)^{1/2}$ ft = 1586 ft (1434 with one bin offset at line intersections)

Desired X_{mute}: ~6000 ft

patch 10 × 48 or 8800 ft × 10 560 ft
number of channels = 480
aspect ratio = 8800 ÷ 10 560 = 0.83
X_{max} = $\tfrac{1}{2}$ × $(8800^2 + 10\,560^2)^{1/2}$ ft = 6873 ft

Fold:

in-line fold = 48 × 220 ft ÷ (2 × 1320 ft) = 4
cross-line fold = 10 ÷ 2 = 5
total fold = 20

Migration Apron:

radius of Fresnel Zone =
   $\tfrac{1}{2}$ × V_{ave} × (target two-way time ÷ f_{dom})$^{1/2}$ =
diffraction energy = 0.58 × target depth =
migration apron = target depth × tan (dip) =
in-line fold taper = [(in-line fold ÷ 2) – 0.5] × SLI =
cross-line fold taper = [(cross-line fold ÷ 2) – 0.5]
× RLI = (FT + FZ) < total migration apron
< (FT + MA) TMA *range* =

Let's Design a 3-D—Summary (metric)

In this example, these are the following known parameters:

existing 2-D data of good quality have	30 fold
steepest dips	20°
shallow markers needed for isochroning have	500 m offsets
target depth	2000 m
target two-way time	1.5 s
basement depth	3000 m
V_{int} immediately above the target horizon	4200 m/s
f_{dom} at the target horizon	50 Hz
f_{max} at the target horizon	70 Hz
lateral target size	300 m

Orthogonal method is to be used.

Desired Fold:

$$\tfrac{1}{2} \text{ up to full 2-D fold} = 15 - 30$$

Bin size:

a) for target size: B = 300 m ÷ 3 = 100 m
c) for alias frequency:
   B = V_{int} ÷ (4 × f_{max} × sin θ)
   = 4200 m/s ÷ (4 × 70 Hz × sin 20°) = 44 m
c) for lateral resolution:
   B = 4200 m/s ÷ (N × 50 Hz)
   = 21 to 42 m
bin size = 30 m × 30 m
RI = 60 m
SI = 60 m

Desired X_{min}: 500 m

RLI = 360 m
SLI = 360 m
X_{min} = $(360^2 + 360^2)^{1/2}$ m = 509 m
(467 m with one bin offset at line intersections)

Desired X_{mute}: 2000 m

patch 8 × 60 or 2880 m × 3600 m
number of channels = 480
aspect ratio = 2880 ÷ 3600 = 0.80
X_{max} = $\tfrac{1}{2}$ × $(2880^2 + 3600^2)^{1/2}$ m = 2305 m

Fold:

in-line fold = 60 × 60 m ÷ (2 × 360 m) = 5
cross-line fold = 8 ÷ 2 = 4
total fold = 20

Migration Apron:

radius of Fresnel Zone = $\frac{1}{2}$ × 2666 m/s × (1.5 s ÷ 50 Hz)$^{1/2}$ = 231 m
diffraction energy = 0.58 × 2000 m = 1155 m
migration apron = 2000 × tan 20° = 728 m
in-line fold taper = [(in-line fold ÷ 2) − 0.5] × SLI = 2.0 × 360 m = 720 m
cross-line fold taper = [(cross-line fold ÷ 2) − 0.5] × RLI = 1.5 × 360 = 540 m
TMA range = 771 m to 1875 m

Let's Design a 3-D – Summary (imperial)

In this example, these are the following known parameters:

existing 2-D data of good quality have	30 fold
steepest dips	20°
shallow markers needed for isochroning have	1500 ft offsets
target depth	6000 ft
target two-way time	1.5 s
basement depth	10 000 ft
V_{int} immediately above the target horizon	14 000 ft/s
f_{dom} at the target horizon	50 Hz
f_{max} at the target horizon	70 Hz
lateral target size	1000 ft

Orthogonal method is to be used.

Desired Fold:

$\frac{1}{2}$ up to full 2-D fold = 15–30

Bin size:

c) for target size: B = 1000 ft ÷ 3 = 333 ft
d) for alias frequency:
B = V_{int} ÷ (4 × f_{max} × sin (dip))
= 14 000 ft/s ÷ (4 × 70 Hz × sin 20°) = 146 ft
c) for lateral resolution: B = V_{int} ÷ (N × f_{dom}) =
14 000 ft/s ÷ (N × 50 Hz)
= 70 to 140 ft

Bin size = 110 ft × 110 ft
RI = 220 ft
SI = 220 ft

Desired X_{min}: ~1500 ft

RLI = 880 ft
SLI = 1320 ft
X_{min} = (880^2 + 1320^2)$^{1/2}$ ft = 1586 ft   (1434 ft with one bin offset at line intersections)

Desired X_{mute}: ~6000 ft

patch 10 × 48 or 8800 ft × 10 560 ft
number of channels = 480
aspect ratio = 8800 ÷ 10 560 = 0.83
X_{max} = $\frac{1}{2}$ × (8800^2 + 10 560^2)$^{1/2}$ ft = 6873 ft

Fold:

in-line fold = 48 × 220 ft ÷ (2 × 1320 ft) = 4
cross-line fold = 10 ÷ 2 = 5
total fold = 20

Migration Apron:

radius of Fresnel Zone =
$\frac{1}{2}$ × 8000 ft/s × (1.5 s ÷ 50 Hz)$^{1/2}$ = 693 ft
diffraction energy = 0.58 × 6000 ft = 3464 ft
migration apron = 6000 ft × tan 20° = 2184 ft
in-line fold taper = [(in-line fold ÷ 2) − 0.5] × SLI =
1.5 × 1320 ft = 1980 ft
cross-line fold taper = [(cross-line fold ÷ 2) − 0.5] ×
RLI = 2 × 880 ft = 1760 ft
TMA range = 2453 ft − 5444 ft

Chapter 3 Quiz

1. Is a wide azimuth survey or a narrow azimuth 3-D survey preferred for AVO work?
2. Based on diffraction energy, how large should the migration apron be?
3. Should the migration apron have the same dimension on all sides of the survey?
4. What are the three zones of the acquisition model, and what type of data is contained in each?

4

Flowcharts, Equations, and Spreadsheets

4.1 3-D DESIGN FLOWCHART

Chapters 2 and 3 developed a fairly standard 3-D design example. The decision-making order used in that example works well for most 3-D design problems. Survey design Table 2.1 should be reviewed as the material in this chapter is considered, because that table summarizes the criteria that can be used to determine the seven key parameters of 3-D design: fold, bin size, X_{min}, X_{max}, migration apron, fold taper, and recording time. The required input parameters for a 3-D survey are summarized in Table 4.1. If one cannot determine all the starting parameters from exploration objectives and previous 2-D seismic data in the area, then some reasonable estimates may suffice.

The flowchart in Table 4.1 should give a reasonable starting design in most cases. Once a starting design has been established, the source type should be selected, and constraints such as surface conditions, costs, and operational considerations should be addressed. For a large survey, operational and cost considerations may determine the final choice of design. There are many layout strategies (Chapter 5) that can be used to good advantage in certain situations. This flowchart is written for orthogonal designs, but it can be adapted in most cases to other design strategies.

4.2 BASIC 3-D EQUATIONS—SQUARE BINS

The basic 3-D equations can be expressed in terms of the following parameters:

SD = source density
NC = number of recording channels in a patch
B = bin size
X_r = in-line half-dimension of the patch
X_s = cross-line half-dimension of the patch
RLI = receiver line interval
SLI = source line interval
A = aspect ratio = $\dfrac{X_s}{X_r}$
U = units factor (10^{-6} for m/km²; 0.03587×10^{-6} for ft/mi²)

Chapters 2 and 3 derived the basic fold equation for 3-D surveys. If the bin area = B^2, then Fold = $SD \times NC \times B^2 \times U$ (midpoints per bin) and

$$SD = \frac{Fold}{NC \times B^2 \times U}. \qquad (4.1)$$

Table 4.1. 3-D design input parameters.

Determine the following parameters from exploration objectives and from existing 2-D seismic data:

fold of good 2-D data
steepest dip
mute for shallow markers needed for isochroning
target depth and mute distance
target two-way time
mute for basement depth and mute distance
V_{int} immediately above the target horizon
f_{dom} at the target horizon
f_{max} at the target horizon
lateral target size
area to be fully imaged
layout method.

58 Flowcharts, Equations, and Spreadsheets

Table 4.2. 3-D design flowchart.

Desired Fold	($\frac{1}{2}$ to 1) X full 2-D fold = _____
Bin size	a) for target size: B = _____
	b) for alias frequency: $B = V_{int} \div (4 \times f_{max} \times \sin \theta)$ = _____
	c) for lateral resolution: $B = V_{int} \div (N \times f_{dom})$,
	(N = 2 to 4) = _____ to _____
	bin size = _____
	RI = _____
	SI = _____
Desired X_{min}: _____	RLI = _____
	SLI = _____
	$X_{min} = (RLI^2 + SLI^2)^{1/2}$ = _____
Desired X_{max}: _____	number of channels in patch = _____
	number of receiver lines = _____
	channels per line = _____
	cross-line dimension = _____
	in-line dimension = _____
	aspect ratio = cross-line dimension of the patch/
	in-line dimension of the patch = _____
	$X_{max} = \frac{1}{2} \times$ (in-line dimension of the patch)^2 +
	(cross-line dimension of the patch)^2 $\frac{1}{2}$ = _____
Fold:	in-line fold = $RLL \div (2 \times SLI)$ = _____
	cross-line fold = $\frac{1}{2}$ NRL = _____
	total fold = _____
Migration Apron:	radius of Fresnel Zone = $\frac{1}{2} \times V_{ave} \times$ (target two-way time $\div f_{dom})^{1/2}$
	= diffraction energy = 0.58 $\times$ target depth =
	Migration apron = target depth $\times$ tan (dip) =
	in-line fold taper = [(in-line fold $\div$ 2) $-$ 0.5] $\times$ SLI =
	cross-line fold taper = [(cross-line fold $\div$ 2) $-$ 0.5] $\times$ RLI =
	($FT + FZ$) < total migration apron < ($FT + MA$)
	TMA =

Example: Fold = 24, NC = 480,
bin size B = 25 m *(82.5 ft)*

$$SD = \frac{24}{480 \times 25^2 \times 10^{-6}} = 80 \text{ sources}/\text{km}^2$$

or

$$SD = \frac{24}{480 \times 82.5^2 \times 0.03587 \times 10^{-6}}$$

$$= 205 \text{ sources}/\text{mi}^2$$

Equation (4.1) is the fundamental relationship governing all 3-D surveys, regardless of the design strategy. There is an assumption in this derivation that recording SD sources per unit area into NC channels gives rise to $SD \times NC$ midpoints inside the unit area. This assumption will be realized in practice only if some of the midpoints arise from sources outside the unit area and receivers inside the unit area under consideration and vice versa. This requirement implies that the receiver patches and the sources fired into each patch must create overlapping areas of midpoints arranged in a regular pattern. Such overlap ensures that there are $SD \times NC$ midpoints in each unit area of the survey. Figure 4.1 shows several overlapping midpoint areas that contribute to the fold in the central area.

If the receiver patches move (roll-along) in such a way that there is insufficient overlap, one observes stripes of lower fold at regular intervals in the survey (see Chapter 9). This striping also occurs if some of the receivers are at offsets $> X_{mute}$ and are not used in processing, e.g., for shallower horizons and smaller mute distances.

The receiver line interval can be calculated as

$$RLI = \frac{X_r \times X_s}{NC \times 2B}. \qquad (4.2)$$

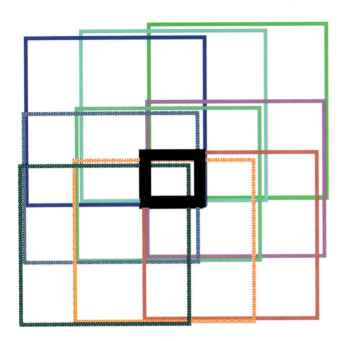

Fig. 4.1. Overlapping midpoint areas.

Example: $X_{max} = 2300$ m *(7500 ft)*, so $X_r = X_s = 3200$ m *(10 600 ft)* and assuming $NC = 480$, $B = 25$ m *(82.5 ft)*, $A \sim 1$

$$RLI = \frac{X_r \times X_s}{NC \times 2B} = \frac{3200^2}{480 \times 2 \times 25} \approx 427 \text{ m}$$

or

$$RLI = \frac{X_r \times X_s}{NC \times 2B} = \frac{10\ 600^2}{480 \times 2 \times 82.5} \approx 1419 \text{ ft.}$$

This expression links *RLI* and *NC*. The assumptions are: a rectangular receiver patch with receivers spaced $2B$ apart and NC receivers in an area with dimensions $X_r \times X_s$; receivers are laid out as parallel lines that are *RLI* apart; and all *NC* channels lie within a useful offset range. These assumptions are not overly rigid. ARCO button patch designs fall into this scheme if one assumes that half the patch is empty. In other words, the button patch design should be calculated using $2 \times NC$ channels instead of *NC*.

The number of receiver lines follows from the cross-line dimension of the patch and the receiver line interval; i.e.,

$$NRL = \frac{X_s}{RLI}. \quad (4.3)$$

From equations (4.2) and (4.3),

$$NRL = \frac{NC \times 2B}{X_r}, \quad (4.4)$$

and combining equations (4.1) and (4.2) results in

$$RLI = \frac{SD \times X_r \times X_s \times B \times U}{2 \times \text{Fold}}. \quad (4.5)$$

The source line interval can be calculated from:

$$SLI = \frac{1}{2B \times SD \times U}. \quad (4.6)$$

Important geometrical considerations are to determine how *NS* sources per square kilometer *(mile)* can be laid out because there must be $1000 \div 2B$ sources per line km *($5280 \div 2B$ sources per line mile)*. The *SLI* value can be thought of as sources per km² *(mi²)* divided by sources per line km *(mi)*.

4.3 BASIC 3-D EQUATIONS—RECTANGULAR BINS

If the CMP bins are rectangular, the preceding bin size B needs to be replaced with:

B_s = bin size in direction of source lines
B_r = bin size in direction of receiver lines

The fundamental design equations are then as follows. First,

$$SD = \frac{\text{Fold}}{NC \times B_s \times B_r \times U}. \quad (4.7)$$

This equation assumes 100% overlapping patterns in each area of the survey, or a regular midpoint density. Second,

$$RLI = \frac{X_r \times X_s}{NC \times 2B_r}. \quad (4.8)$$

This equation assumes a rectangular receiver patch with receiver stations spaced $2B_r$ apart and NC receivers in area $X_r \times X_s$. Other key equations are

$$NRL = \frac{NC \times 2B_r}{X_r}, \quad (4.9)$$

$$RLI = \frac{SD \times X_r \times X_s \times B_r \times U}{\text{Fold}}, \quad (4.10)$$

60   Flowcharts, Equations, and Spreadsheets

$$SLI = \frac{1}{2B_s \times SD \times U}.$$ (4.11)

The above equations work well in cases where source lines and receiver lines are perpendicular, e.g., orthogonal, brick, and *Flexi-Bin® designs. In other designs, such as nonorthogonal or zig-zag, one must pay attention to how the source and source line intervals are defined. The best method is to base the direction of measurement on the basis of the bin direction.

4.4 BASIC STEPS IN 3-D LAYOUT— FIVE-STEP METHOD

From the basic 3-D equations and the Survey Design Decision Table (Table 2.1), one can now implement a five-step process to design a 3-D survey. This approach is similar to the 3-D Design Flowchart presented in Table 4.2, but it involves a spreadsheet to help choose some parameters.

1. Based on geologic modeling considerations, one should decide:
   a. full fold survey size
   b. maximum frequency desired
   c. fold
   d. X_{min}
   e. X_{max}
   f. bin size B
   g. migration apron, and how much overlap can be tolerated between migration apron and fold taper
   h. maximum recording time
2. Create a spreadsheet with columns calculated according to the format in Table 4.3. Next, choose the following four input parameters: fold, bin size, X_r and X_s.
   - **Col 1 (SLI)** Enter values of source line interval from below X_{min} to $2 \times X_{min}$ at steps of the group interval ($2 \times B$), (e.g., 200 m to 700 m in steps of 50 m; or 660 ft to 2310 ft in steps of 165 ft).
   - **Col 2 (SD)** = 1 ÷ (2 B × **Col 1** × U)   see equation (4.6)
   - **Col 3 (NC)** = fold ÷ (**Col 2** × B^2 × U)   see equation (4.1)
   - **Col 4 (NRL)** = **Col 3** × 2B ÷ X_r see equation (4.4)
   - **Col 5 (RLI)** = X_s ÷ **Col 4**   see equation (4.3)
   - **Col 6 (X_{min})** = [(**Col 1**)^2 + (**Col 5**)^2]$^-$; see equation (2.42)

   Choose the spreadsheet row for a solution where the combination of SLI and RLI nearly satisfy the

Table 4.3a. Spreadsheet for evaluating 3-D designs (metric).

| Fold: | 40 | | X_r: | 2000 m |
| Bin Size: | 25 m | | X_s: | 1500 m |

For integer NRL, choose fold × SLI ÷ X_r = integer; this equals the cross-line fold

SLI	SD	NC	NRL	RLI	X_{min} orthogonal	X_{min} brick
200	100.0	640	8.0	429	473	293
250	80.0	800	10.0	333	417	300
300	66.7	960	12.0	273	405	330
350	57.1	1120	14.0	231	419	369
400	50.0	1280	16.0	200	447	412
450	44.4	1440	18.0	176	483	459
500	40.0	1600	20.0	158	524	506
550	36.4	1760	22.0	143	568	555
600	33.3	1920	24.0	130	614	604
650	30.8	2080	26.0	120	661	653
700	28.6	2240	28.0	111	709	702

Note: Distances are in m and units of SD are shots/km^2
This table assumes coincident source and receiver positions at line intersections.

*Flexi-Bin® is a registered trademark of Geophysical Exploration & Development Corporation, Calgary, Alberta, Canada.

Table 4.3b. Spreadsheet for evaluating 3-D designs (imperial).

| Fold: | 40 | | X_r: | 6600 ft |
| Bin Size: | 82.5 ft | | X_s: | 5500 ft |

For integer NRL, choose Fold × SLI ÷ X_r = integer; this equals the cross-line fold

SLI	SD	NC	NRL	RLI	X_{min} orthogonal	X_{min} brick
660	256.0	640	8.0	1571	1704	1026
825	204.8	800	10.0	1222	1475	1027
990	170.7	960	12.0	1000	1407	1109
1155	146.3	1120	14.0	846	1432	1230
1320	128.0	1280	16.0	733	1510	1370
1485	113.8	1440	18.0	647	1620	1520
1650	102.4	1600	20.0	579	1749	1675
1815	93.1	1760	22.0	524	1889	1834
1980	85.3	1920	24.0	478	2037	1994
2145	78.8	2080	26.0	440	2190	2156
2310	73.1	2240	28.0	407	2346	2319

Note: Distances are in ft and units of SD are shots/mi^2.
This table assumes coincident source and receiver positions at line intersections.

near-offset (X_{min}) constraint. Select a receiver line interval that is smaller than necessary for the required X_{min} and that is a multiple of the station interval. Adjust X_r and X_s within reasonable bounds for the desired stacking offset to create a patch with an even number of receiver lines. To achieve an even number of receiver lines, the expression, fold $\times$ SLI $\div$ X_r must be an integer (this quantity is the cross-line fold).
3. Choose a layout strategy such as orthogonal line, depending on practical considerations such as access and available equipment. Add exclusion zones and move source and receiver lines to accommodate real-life features. Move and add sources where necessary to preserve fold, offset, and azimuth mix. Determine migration and fold taper needs.
4. Model the 3-D survey for fold, offset and azimuth variations, as well as edge management. Numerous quality control displays should be made to assure a high-quality acquisition.
5. Prepare preplots and script files for the field operations.

Other methods are possible, but equation (4.1) [$NS =$ Fold $\div$ ($NC \times B^2 \times U$)] is fundamental to any 3-D layout strategy.

4.5 GRAPHICAL APPROACH

The requirements for a graphical approach to 3-D design are as follows:

Fold = 25, B = 25 m,
$\quad X_r = X_s = 5000$ m with $X_{max} = 3500$ m, $A = 1$

or

Fold = 25, B = 82.5 ft,
$\quad X_r = X_s = 16\ 500$ ft with $X_{max} = 11\ 550$ ft, $A = 1$

Two equations limit the source density SD. The first equation relates SD to RLI:

$$SD = \frac{2 \times \text{Fold} \times RLI}{X_s \times X_r \times B \times U} = 0.08 \times RLI$$

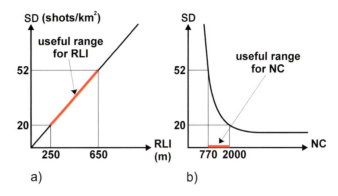

Fig. 4.2. Graphical determination of acquisistion parameters.

[See equation (4.9) and graph U of Figure 4.2a]

or = 0.0621 × RLI.

The second equation relates SD to NC:

$$SD = \frac{\text{Fold}}{NC \times B^2 \times U} = \frac{40\ 000}{NC}$$

[See equation (4.1) and graph of Figure 4.2b]

or $$NS = \frac{102\ 400}{NC}.$$

Assume that 700 m *(2310 ft)* is the largest minimum acceptable offset X_{min} in any bin. This requirement can be achieved with RLI in the range of 250 to 600 m *(825 to 1980 ft)*.

To achieve these parameters (Fold = 25, B = 25 m, $X_r = X_s = 5000$ m, RLI = 250-650 m *or* Fold = 25, B = 82.5 ft, $X_r = X_s = 16\ 500$ ft, RLI = 825 to 1980 ft), NS has to be in the range of 20 to 52 sources / km² *(51 to 123 sources/mi²)*, and NC has to be in the range of 770 to 2000 *(833 to 2008)* channels.

In this case SLI = 400 to 1000 m *(1320 to 3300 ft)* [from equation (4.11)]. To optimize X_{min}, SLI should be as close to the receiver line interval as possible. If RLI = SLI = 500 m, then X_{min} is 700 m/*(1760 ft, then X_{min} is 2489 ft)*, SD = 40 sources / km² *(96 sources/mi²)*, and NC = 1000 *(1066)* channels.

4.6 STANDARDIZED SPREADSHEETS

A simple way to review numerous design choices is to create standardized spreadsheets that are based on fold and bin size as follows:

Bin Size   30 m   20 fold

Design	RI (m)	SI (m)	RLI m)	SLI (m)	Channels	Lines	Fold	X_{min} (m)	X_{max} (m)	Aspect ratio
1	60	60	120	120	160	8	20	127	768	0.80
2	60	60	120	180	240	10	20	175	937	0.83
3	60	60	120	240	320	10	20	228	1132	0.63
4	60	60	180	180	240	8	20	212	1152	0.80
5	60	60	180	240	320	10	20	258	1316	0.94
6	60	60	180	300	400	10	20	309	1500	0.75
7	60	60	240	240	320	8	20	297	1537	0.80
8	60	60	240	300	400	10	20	342	1697	1.00
9	60	60	240	360	480	10	20	391	1874	0.83
10	60	60	300	300	400	8	20	382	1921	0.80
11	60	60	300	360	480	8	20	426	2163	0.67
12	60	60	300	420	560	8	20	474	2252	0.89
13	**60**	**60**	**360**	**360**	**480**	**8**	**20**	**467**	**2305**	**0.80**
14	60	60	360	480	640	10	20	558	2632	0.94
15	60	60	480	480	640	8	20	636	3073	0.80
16	60	60	480	600	800	10	20	726	3394	1.00
17	60	60	480	720	960	10	20	824	3749	0.83
18	60	60	600	600	800	8	20	806	3842	0.80
19	60	60	720	720	960	8	20	976	4610	0.80
20	60	60	960	960	1280	8	20	1315	6147	0.80

The solutions that are presented are limited to typical designs used for land 3-D surveys with an aspect ratio <1.0. In a spreadsheet form, one can make changes to suit each prospect environment. Such spreadsheets allow a quick review of several solutions with the associated minimum and maximum offsets as well as the aspect ratios. The tables in Section 4.6 assume that the lines were shifted by one bin size from their coincident positions at the line intersections. Note that the receiver line interval can be changed without affecting the fold; however, the minimum and maximum offsets, as well as the aspect ratios, would be changed. Design 13 is one of the possible solutions for the "Let's Design a 3-D" exercises.

The same table can be converted to imperial units:

Bin Size   110 ft   20 fold

Design	RI (ft)	SI (ft)	RLI (ft)	SLI (ft)	Channel	Lines	Fold	X_{min} (ft)	X_{max} (ft)	Aspect ratio
1	220	220	440	440	160	8	20	467	2817	0.80
2	220	220	440	660	240	10	20	641	3437	0.83
3	220	220	440	880	320	10	20	838	4151	0.63
4	220	220	660	660	240	8	20	778	4226	0.80
5	220	220	660	880	320	10	20	946	4825	0.94
6	220	220	660	1100	400	10	20	1133	5500	0.75
7	220	220	880	880	320	8	20	1089	5635	0.80
8	220	220	880	1100	400	10	20	1254	6223	1.00
9	**220**	**220**	**880**	**1320**	**480**	**10**	**20**	**1434**	**6873**	**0.83**
10	220	220	1100	1100	400	8	20	1400	7043	0.80
11	220	220	1100	1320	480	8	20	1563	7932	0.67
12	220	220	1100	1540	560	10	20	1739	8258	0.89
13	220	220	1320	1320	480	8	20	1711	8452	0.80
14	220	220	1320	1760	640	10	20	2046	9650	0.94
15	220	220	1760	1760	640	8	20	2333	11269	0.80
16	220	220	1760	2200	800	10	20	2663	12445	1.00
17	220	220	1760	2640	960	10	20	3020	13746	0.83
18	220	220	2200	2200	800	8	20	2956	14087	0.80
19	220	220	2640	2640	960	8	20	3578	16904	0.80
20	220	220	3520	3520	1280	8	20	4822	22539	0.80

Design 9 is one of the possible answers for the "Let's Design a 3-D" exercises.

The following pages summarize numerous solutions for orthogonal 3-D surveys. These solutions allow the selection of suitable 3-D designs for a variety of different situations.

Bin Size   25 m   12 fold

Design	RI (m)	SI (m)	RLI (m)	SLI (m)	Channels	Lines	Fold	X_{min} (m)	X_{max} (m)	Aspect ratio
1	50	50	100	100	96	6	12	106	500	0.75
2	50	50	100	150	144	8	12	146	602	0.89
3	50	50	100	200	192	8	12	190	721	0.67
4	50	50	150	150	144	6	12	177	750	0.75
5	50	50	150	200	192	8	12	215	849	1.00
6	50	50	150	250	240	8	12	257	960	0.80
7	50	50	200	200	192	6	12	247	1000	0.75
8	50	50	200	250	240	6	12	285	1166	0.60
9	50	50	200	300	288	8	12	326	1204	0.89
10	50	50	250	250	240	6	12	318	1250	0.75
11	50	50	250	300	288	6	12	355	1415	0.63
12	50	50	250	350	336	8	12	395	1450	0.95
13	50	50	300	300	288	6	12	389	1500	0.75
14	50	50	300	400	384	8	12	465	1697	1.00
15	50	50	400	400	384	6	12	530	2000	0.75
16	50	50	400	500	480	6	12	605	2332	0.60
17	50	50	400	600	576	8	12	686	2408	0.89
18	50	50	500	500	480	6	12	672	2500	0.75
19	50	50	600	600	576	6	12	813	3000	0.75
20	50	50	800	800	768	6	12	1096	4000	0.75

64 Flowcharts, Equations, and Spreadsheets

Bin Size   25 m   16 fold

Design	RI (m)	SI (m)	RLI (m)	SLI (m)	Channels	Lines	Fold	X_{min} (m)	X_{max} (m)	Aspect ratio
1	50	50	100	100	128	8	16	106	566	1.00
2	50	50	100	150	192	8	16	146	721	0.67
3	50	50	100	200	256	8	16	190	894	0.50
4	50	50	150	150	192	8	16	177	849	1.00
5	50	50	150	200	256	8	16	215	1000	0.75
6	50	50	150	250	320	8	16	257	1166	0.60
7	50	50	200	200	256	8	16	247	1131	1.00
8	50	50	200	250	320	8	16	285	1281	0.80
9	50	50	200	300	384	8	16	326	1442	0.67
10	50	50	250	250	320	8	16	318	1414	1.00
11	50	50	250	300	384	8	16	355	1562	0.83
12	50	50	250	350	448	8	16	395	1720	0.71
13	50	50	300	300	384	8	16	389	1697	1.00
14	50	50	300	400	512	8	16	465	2000	0.75
15	50	50	400	400	512	8	16	530	2263	1.00
16	50	50	400	500	640	8	16	605	2561	0.80
17	50	50	400	600	768	8	16	686	2884	0.67
18	50	50	500	500	640	8	16	672	2828	1.00
19	50	50	600	600	768	8	16	813	3394	1.00
20	50	50	800	800	1024	8	16	1096	4525	1.00

Bin Size   25 m   20 fold

Design	RI (m)	SI (m)	RLI (m)	SLI (m)	Channels	Lines	Fold	X_{min} (m)	X_{max} (m)	Aspect ratio
1	50	50	100	100	160	8	20	106	640	0.80
2	50	50	100	150	240	10	20	146	781	0.83
3	50	50	100	200	320	10	20	190	943	0.63
4	50	50	150	150	240	8	20	177	960	0.80
5	50	50	150	200	320	10	20	215	1097	0.94
6	50	50	150	250	400	10	20	257	1250	0.75
7	50	50	200	200	320	8	20	247	1281	0.80
8	50	50	200	250	400	10	20	285	1414	1.00
9	50	50	200	300	480	10	20	326	1562	0.83
10	50	50	250	250	400	8	20	318	1601	0.80
11	50	50	250	300	480	8	20	355	1803	0.67
12	50	50	250	350	560	10	20	395	1877	0.89
13	50	50	300	300	480	8	20	389	1921	0.80
14	50	50	300	400	640	10	20	465	2193	0.94
15	50	50	400	400	640	8	20	530	2561	0.80
16	50	50	400	500	800	10	20	605	2828	1.00
17	50	50	400	600	960	10	20	686	3124	0.83
18	50	50	500	500	800	8	20	672	3202	0.80
19	50	50	600	600	960	8	20	813	3842	0.80
20	50	50	800	800	1280	8	20	1096	5122	0.80

4.6 Standardized Spreadsheets

Bin Size   25 m   24 fold

Design	RI (m)	SI (m)	RLI (m)	SLI (m)	Channels	Lines	Fold	X_{min} (m)	X_{max} (m)	Aspect ratio
1	50	50	100	100	192	8	24	106	721	0.67
2	50	50	100	150	288	12	24	146	849	1.00
3	50	50	100	200	384	12	24	190	1000	0.75
4	50	50	150	150	288	8	24	177	1082	0.67
5	50	50	150	200	384	8	24	215	1342	0.50
6	50	50	150	250	480	12	24	257	1345	0.90
7	50	50	200	200	384	8	24	247	1442	0.67
8	50	50	200	250	480	8	24	285	1700	0.53
9	50	50	200	300	576	12	24	326	1697	1.00
10	50	50	250	250	480	8	24	318	1803	0.67
11	50	50	250	300	576	8	24	355	2059	0.56
12	50	50	250	350	672	8	24	395	2326	0.48
13	50	50	300	300	576	8	24	389	2163	0.67
14	50	50	300	400	768	8	24	465	2683	0.50
15	50	50	400	400	768	8	24	530	2884	0.67
16	50	50	400	500	960	8	24	605	3400	0.53
17	50	50	400	600	1152	12	24	686	3394	1.00
18	50	50	500	500	960	8	24	672	3606	0.67
19	50	50	600	600	1152	8	24	813	4327	0.67
20	50	50	800	800	1536	8	24	1096	5769	0.67

Bin Size   25 m   30 fold

Design	RI (m)	SI (m)	RLI (m)	SLI (m)	Channels	Lines	Fold	X_{min} (m)	X_{max} (m)	Aspect ratio
1	50	50	100	100	240	10	30	106	781	0.83
2	50	50	100	150	360	12	30	146	960	0.80
3	50	50	100	200	480	12	30	190	1166	0.60
4	50	50	150	150	360	10	30	177	1172	0.83
5	50	50	150	200	480	12	30	215	1345	0.90
6	50	50	150	250	600	12	30	257	1540	0.72
7	50	50	200	200	480	10	30	247	1562	0.83
8	50	50	200	250	600	12	30	285	1733	0.96
9	50	50	200	300	720	12	30	326	1921	0.80
10	50	50	250	250	600	10	30	318	1953	0.83
11	50	50	250	300	720	12	30	355	2121	1.00
12	50	50	250	350	840	12	30	395	2305	0.86
13	50	50	300	300	720	10	30	389	2343	0.83
14	50	50	300	400	960	12	30	465	2691	0.90
15	50	50	400	400	960	10	30	530	3124	0.83
16	50	50	400	500	1200	12	30	605	3466	0.96
17	50	50	400	600	1440	12	30	686	3842	0.80
18	50	50	500	500	1200	10	30	672	3905	0.83
19	50	50	600	600	1440	10	30	813	4686	0.83
20	50	50	800	800	1920	10	30	1096	6248	0.83

66 Flowcharts, Equations, and Spreadsheets

Bin Size 25 m 40 fold

Design	RI (m)	SI (m)	RLI (m)	SLI (m)	Channels	Lines	Fold	X_{min} (m)	X_{max} (m)	Aspect ratio
1	50	50	100	100	320	10	40	106	943	0.63
2	50	50	100	150	480	10	40	146	1300	0.42
3	50	50	100	200	640	16	40	190	1281	0.80
4	50	50	150	150	480	10	40	177	1415	0.63
5	50	50	150	200	640	10	40	215	1767	0.47
6	50	50	150	250	800	16	40	257	1733	0.96
7	50	50	200	200	640	10	40	247	1887	0.63
8	50	50	200	250	800	10	40	285	2236	0.50
9	50	50	200	300	960	10	40	326	2600	0.42
10	50	50	250	250	800	10	40	318	2358	0.63
11	50	50	250	300	960	10	40	355	2706	0.52
12	50	50	250	350	1120	10	40	395	3066	0.45
13	50	50	300	300	960	10	40	389	2830	0.63
14	50	50	300	400	1280	10	40	465	3534	0.47
15	50	50	400	400	1280	10	40	530	3774	0.63
16	50	50	400	500	1600	10	40	605	4472	0.50
17	50	50	400	600	1920	10	40	686	5200	0.42
18	50	50	500	500	1600	10	40	672	4717	0.63
19	50	50	600	600	1920	10	40	813	5660	0.63
20	50	50	800	800	2560	10	40	1096	7547	0.63

Bin Size 110 ft 12 fold

Design	RI (ft)	SI (ft)	RLI (ft)	SLI (ft)	Channels	Lines	Fold	X_{min} (ft)	X_{max} (ft)	Aspect ratio
1	220	220	440	440	96	6	12	467	2200	0.75
2	220	220	440	660	144	8	12	641	2649	0.89
3	220	220	440	880	192	8	12	838	3173	0.67
4	220	220	660	660	144	6	12	778	3300	0.75
5	220	220	660	880	192	8	12	946	3734	1.00
6	220	220	660	1100	240	8	12	1133	4226	0.80
7	220	220	880	880	192	6	12	1089	4400	0.75
8	220	220	880	1100	240	6	12	1254	5131	0.60
9	220	220	880	1320	288	8	12	1434	5298	0.56
10	220	220	1100	1100	240	6	12	1400	5500	0.75
11	220	220	1100	1320	288	6	12	1563	6226	0.63
12	220	220	1100	1540	336	8	12	1739	6380	0.95
13	220	220	1320	1320	288	6	12	1711	6600	0.75
14	220	220	1320	1760	384	8	12	2046	7467	1.00
15	220	220	1760	1760	384	6	12	2333	8800	0.75
16	220	220	1760	2200	480	6	12	2663	10262	0.60
17	220	220	1760	2640	576	8	12	3020	10597	0.89
18	220	220	2200	2200	480	6	12	2956	11000	0.75
19	220	220	2640	2640	576	6	12	3578	13200	0.75
20	220	220	3520	3520	768	6	12	4822	17600	0.75

4.6 Standardized Spreadsheets

Bin Size   110 ft   16 fold

Design	RI (ft)	SI (ft)	RLI (ft)	SLI (ft)	Channels	Lines	Fold	X$_{min}$ (ft)	X$_{max}$ (ft)	Aspect ratio
1	220	220	440	440	128	8	16	467	2489	1.00
2	220	220	440	660	192	8	16	641	3173	0.67
3	220	220	440	880	256	8	16	838	3935	0.50
4	220	220	660	660	192	8	16	778	3734	1.00
5	220	220	660	880	256	8	16	946	4400	0.75
6	220	220	660	1100	320	8	16	1133	5131	0.60
7	220	220	880	880	256	8	16	1089	4978	1.00
8	220	220	880	1100	320	8	16	1254	5635	0.80
9	220	220	880	1320	384	8	16	1434	6346	0.67
10	220	220	1100	1100	320	8	16	1400	6223	1.00
11	220	220	1100	1320	384	8	16	1563	6873	0.83
12	220	220	1100	1540	448	8	16	1739	7570	0.71
13	220	220	1320	1320	384	8	16	1711	7467	1.00
14	220	220	1320	1760	512	8	16	2046	8800	0.75
15	220	220	1760	1760	512	8	16	2333	9956	1.00
16	220	220	1760	2200	640	8	16	2663	11269	0.80
17	220	220	1760	2640	768	8	16	3020	12692	0.67
18	220	220	2200	2200	640	8	16	2956	12445	1.00
19	220	220	2640	2640	768	8	16	3578	14934	1.00
20	220	220	3520	3520	1024	8	16	4822	19912	1.00

Bin Size   110 ft   20 fold

Design	RI (ft)	SI (ft)	RLI (ft)	SLI (ft)	Channels	Lines	Fold	X$_{min}$ (ft)	X$_{max}$ (ft)	Aspect ratio
1	220	220	440	440	160	8	20	467	2817	0.80
2	220	220	440	660	240	10	20	641	3437	0.83
3	220	220	440	880	320	10	20	838	4151	0.63
4	220	220	660	660	240	8	20	778	4226	0.80
5	220	220	660	880	320	10	20	946	4825	0.94
6	220	220	660	1100	400	10	20	1133	5500	0.75
7	220	220	880	880	320	8	20	1089	5635	0.80
8	220	220	880	1100	400	10	20	1254	6223	1.00
9	220	220	880	1320	480	10	20	1434	6873	0.83
10	220	220	1100	1100	400	8	20	1400	7043	0.80
11	220	220	1100	1320	480	8	20	1563	7932	0.67
12	220	220	1100	1540	560	10	20	1739	8258	0.89
13	220	220	1320	1320	480	8	20	1711	8452	0.80
14	220	220	1320	1760	640	10	20	2046	9650	0.94
15	220	220	1760	1760	640	8	20	2333	11269	0.80
16	220	220	1760	2200	800	10	20	2663	12445	1.00
17	220	220	1760	2640	960	10	20	3020	13746	0.83
18	220	220	2200	2200	800	8	20	2956	14087	0.80
19	220	220	2640	2640	960	8	20	3578	16904	0.80
20	220	220	3520	3520	1280	8	20	4822	22539	0.80

68 Flowcharts, Equations, and Spreadsheets

Bin Size   110 ft   24 fold

Design	RI (ft)	SI (ft)	RLI (ft)	SLI (ft)	Channels	Lines	Fold	X_{min} (ft)	X_{max} (ft)	Aspect ratio
1	220	220	440	440	192	8	24	467	3173	0.67
2	220	220	440	660	288	12	24	641	3734	1.00
3	220	220	440	880	384	12	24	838	4400	0.75
4	220	220	660	660	288	8	24	778	4759	0.67
5	220	220	660	880	384	8	24	946	5903	0.50
6	220	220	660	1100	480	12	24	1133	5920	0.90
7	220	220	880	880	384	8	24	1089	6346	0.67
8	220	220	880	1100	480	8	24	1254	7480	0.53
9	220	220	880	1320	576	12	24	1434	7467	1.00
10	220	220	1100	1100	480	8	24	1400	7932	0.67
11	220	220	1100	1320	576	8	24	1563	9060	0.56
12	220	220	1100	1540	672	8	24	1739	10234	0.48
13	220	220	1320	1320	576	8	24	1711	9519	0.67
14	220	220	1320	1760	768	8	24	2046	11806	0.50
15	220	220	1760	1760	768	8	24	2333	12692	0.67
16	220	220	1760	2200	960	8	24	2663	14960	0.53
17	220	220	1760	2640	1152	12	24	3020	14934	1.00
18	220	220	2200	2200	960	8	24	2956	15864	0.67
19	220	220	2640	2640	1152	8	24	3578	19037	0.67
20	220	220	3520	3520	1536	8	24	4822	25383	0.67

Bin Size   110 ft   30 fold

Design	RI (ft)	SI (ft)	RLI (ft)	SLI (ft)	Channels	Lines	Fold	X_{min} (ft)	X_{max} (ft)	Aspect ratio
1	220	220	440	440	240	10	30	622	3213	0.78
2	220	220	440	660	360	12	30	793	4004	0.76
3	220	220	440	880	480	12	30	984	4925	0.56
4	220	220	660	660	360	10	30	933	4862	0.77
5	220	220	660	880	480	12	30	1100	5620	0.85
6	220	220	660	1100	600	12	30	1283	6498	0.67
7	220	220	880	880	480	10	30	1245	6512	0.77
8	220	220	880	1100	600	12	30	1409	7244	0.90
9	220	220	880	1320	720	12	30	1586	8096	0.75
10	220	220	1100	1100	600	10	30	1556	8162	0.76
11	220	220	1100	1320	720	12	30	1718	8873	0.93
12	220	220	1100	1540	840	12	30	1893	9706	0.80
13	220	220	1320	1320	720	10	30	1867	9812	0.76
14	220	220	1320	1760	960	12	30	2046	11839	0.90
15	220	220	1760	1760	960	10	30	2333	13746	0.83
16	220	220	1760	2200	1200	12	30	2663	15248	0.96
17	220	220	1760	2640	1440	12	30	3020	16904	0.80
18	220	220	2200	2200	1200	10	30	2956	17183	0.83
19	220	220	2640	2640	1440	10	30	3578	20619	0.83
20	220	220	3520	3520	1920	10	30	4822	27492	0.83

Bin Size   110 ft   40 fold

Design	RI (ft)	SI (ft)	RLI (ft)	SLI (ft)	Channels	Lines	Fold	X_{min} (ft)	X_{max} (ft)	Aspect ratio
1	220	220	440	440	320	10	40	622	3943	0.58
2	220	220	440	660	480	16	40	793	4590	1.03
3	220	220	440	880	640	16	40	984	5412	0.77
4	220	220	660	660	480	10	40	933	5962	0.57
5	220	220	660	880	640	10	40	1100	7540	0.43
6	220	220	660	1100	800	16	40	1283	7318	0.92
7	220	220	880	880	640	10	40	1245	7982	0.57
8	220	220	880	1100	800	10	40	1409	9550	0.46
9	220	220	880	1320	960	16	40	1586	9256	1.02
10	220	220	1100	1100	800	10	40	1556	10001	0.57
11	220	220	1100	1320	960	10	40	1718	11563	0.47
12	220	220	1100	1540	1120	10	40	1893	13175	0.41
13	220	220	1320	1320	960	10	40	1867	12020	0.57
14	220	220	1320	1760	1280	10	40	2046	15550	0.47
15	220	220	1760	1760	1280	10	40	2333	16604	0.63
16	220	220	1760	2200	1600	10	40	2663	19677	0.50
17	220	220	1760	2640	1920	10	40	3020	22880	0.42
18	220	220	2200	2200	1600	10	40	2956	20755	0.63
19	220	220	2640	2640	1920	10	40	3578	24906	0.63
20	220	220	3520	3520	2560	10	40	4822	33208	0.63

4.7 SPREADSHEET FOR A 3-D DESIGN FLOWCHART

Correctly estimating the costs associated with a seismic 3-D survey can be a lengthy procedure because many factors must be taken into account. When working in one locality or under certain conditions with similar parameters from one survey to the next, one may be able to develop a costing model that includes the major factors that influence cost variations. Such models can be developed in a graph form, as simple equations, or as a spreadsheet. Often even before bids are requested from acquisition contractors, management may require a cost estimate to make initial decisions on factors such as the size of the survey.

For such purposes, the spreadsheet example on the following pages provides insight into the biggest cost factors. Such a spreadsheet should be adapted to local conditions to provide the user with the fastest method of examining the factors that control costs.

This particular spreadsheet is designed for use with an orthogonal survey. The basic parameters from the survey design decision table, along with some geological input parameters, are summarized at the top. The first page describes the various input parameters, including some economic criteria; the geometry calculations are based on starting with a certain number of channels. The second page summarizes the acquisition effort and includes the profitability table. The third page calculates the various geometries that are important to know when evaluating various designs. The most important factors here are to get close to the desired fold and to have integer values for inline and crossline fold. X_{min} and X_{max} need to be examined in light of the requirements. The fourth page provides the user with a cost estimate that depends on the variables, which need to be considered for the area in question. The last page is simply a summary of the parameters that are important when requesting bids from acquisition contractors. These numbers are repeated from the earlier pages.

Project:	SEG 3-D		**Client:**	ABC Oil Company
Location:	TEXAS		**GEDCO file #:**	
			9/1/99	

PROSPECT INFORMATION

UNITS	units	M	M (metric) or I (imperial)		
	fold of good 2-D data	20	at	10	m trace spacing
DIP	steepest dips	5.0	degrees	WSW	
SM	mute for shallow markers needed for isochroning	600	m		
DEPTH	target depth	1900	m		
TWT	target two-way time	1.4	s		
BASE	basement depth	2500	m		
V_{int}	V_{int} immediately above the target horizon	3000	m/s		
f_{dom}	f_{dom} at the target horizon	40	Hz		
f_{max}	f_{max} at the target horizon	70	Hz		
	lateral target size	200	m		
	area to be fully imaged	80	km²		
	layout method	orthogonal			

		desired		calculated	
	fold	20		30	
	bin size	25	m	25	m
	X_{min}	600	m	361	m
	X_{max}	2000	m	2052	m
	total migration apron range			754 m to 1697 m	

INPUT

	shape	rectangle		
	receiver line direction	EW		
	source line direction	NS		
RI	receiver interval	50	m	
SI	source interval	50	m	
PB_r	processing bin in receiver direction	25	m	
PB_s	processing bin in source direction	25	m	
RLI	receiver line interval	200	m	affects cross-line offset and patch size
SLI	source line interval	300	m	affects in-line fold, fold
LRL	length of receiver lines	8.000	km	
LSL	length of source lines	10.000	km	
NC	number of channels	720		affects fold, patch size try 720 channels
LINES/REC	patch	12 × 60		
	receiver array	9	over 5	m circle
	source array	1	over 50	m
IROLL	roll on/off stations	Yes		
X ROLL	roll on/off lines	Yes		
SWATH	swath width (max = NRL/2)	3		check for fold striping
DROP	number of consecutive shots which may be dropped	5 out of 80 for access reasons only—not on adjoining source lines except along the edges of the survey		
X OFFSET1	min. distance of any shot from edge of patch	800	m	
NOCUT	% of lines not needing line cuts	30	%	⎫ should total 100%
CUT	% of existing cut lines	25	%	⎪ should total 100%
CATCUT	% of cat cut lines	10	%	⎬ should total 100%
HAND	% of hand-cut lines	35	%	⎭ should total 100%
CROP	% of line km on farm land with crop	20	%	
	% of acceptable dead traces	2.0	%	
SHOT/DAY	estimated shotpoints per day	250		

4.7 Spreadsheet for 3-D Design Flowchart

Project:	SEG 3-D		Client:	ABC Oil Company
Location:	TEXAS		GEDCO file #:	
			9/1/99	

SOURCE PARAMETERS

dynamite

HOLES	number of holes	3	
HOLEDEPTH	hole depth	15	m
CHARGE	charge size per hole	1	kg

vibroseis

	number of vibrators	4	
	number of sweeps	8	
	sweep length	12	s
	pad time	96	s
	sweep frequency—start	10	Hz
	sweep frequency - end	90	Hz
	sweep rate	6.7	Hz/s
	sweep type	3	dB/octave
	source interval	50	m
	max move-up for each sweep	6.25	m

other sources
UTM coordinates
X of SW corner
X of NE corner
Y of SW corner
Y of NE corner
central meridian
in-line angle - rec.

PROFITABILITY TABLE

		without 3-D	with 3-D	0% change
Psource	probability of source	90%	90%	0%
Pmigration	probability of migration/hydrocarbons	80%	80%	0%
Preservoir	probability of reservoir/porosity	70%	80%	14%
Ptrap	probability of seal/trap	30%	40%	33%
NPVsuccess	net present value of successful well	$8,000,000	$8,000,000	
NPVfailure	net present value of dry well	($1,500,000)	($1,500,000)	
Pes	probability of economic success	15%	23%	52%
Pef	probability of economic failure	85%	77%	9%
EMV	expected monetary value	($63,600)	$688,800	
VOI	Value of Information (e.g., 2-D, 3-D, interpretations)		($752,400)	
	Required success ratio for single well	16%	22%	39%

Project:	SEG 3-D		Client:	ABC Oil Company		
Location:	TEXAS		GEDCO file #:			
			9/1/99			

CALCULATED

NBR/S	natural bin dimensions	25	×	25	m	
SBR/S	sub-bin dimensions	25	×	25	m	
	RLI/SI	4.00				
	SLI/RI	6.00				
NBINS	number of processing bins	128000				
NTRACE	number of recorded traces	4003920				
		full fold	at offset	sub-bin		
IFOLD	in-line fold	5.0	5.0	5.0		
XFOLD	cross-line fold	6.0	6.0	6.0		
FOLD	total fold	30.0	30.0	30.0		
	desired 3-D fold according to Krey	23.2				
	maximum fold assuming all receivers live	1333				
IROLLS	in-line rolls	26				
XROLLS	cross-line rolls	15				
ROLLS	total number of rolls	431				
	in-line templates	27				
	cross-line templates	16				
TEMPLATES	total number of template positions	432				
V_{ave}	average velocity to the target zone	2714	m/s			
FZ	Fresnel zone radius before migration	254	m			
DIFF	apron to capture ~95% of diffraction energy	1097	m			
MA	migration aperture	166	m			
TAPER	fold taper @ 20% of target depth	380	m			
ITAPER	TAPER-in-line	600	m			
XTAPER	TAPER-cross-line	500	m			
	fold rate - in-line	15	per *SLI*			
	fold rate - cross-line	12	per *RLI*			
	periodicity - in-line	0	bins or	0.0	source lines	
	periodicity - cross-line	24	bins or	3.0	receiver lines	
TMA	total migration aperture range	754	m to	1697	m	

4.7 Spreadsheet for 3-D Design Flowchart

Fal	aliasing frequency before migration	344	Hz	
Falm	aliasing frequency after migration	343	Hz	
LRES	lateral resolution	38	m	bin size should be smaller than this
VRES	vertical resolution	38	m	
kN_r	Nyquist wavenumber (receiver direction)	0.800	1/m	
	apparent velocites smaller than	88	m/s	will be suppressed by receiver array
kN_s	Nyquist wavenumber (source direction)	0.000	1/m	
	apparent velocites smaller than	0	m/s	no source array
SIZE	size of survey	80.0	km^2	
NETSIZE	size of survey net of total migration aperture	58.4	km^2	taper as 20% of depth plus Fresnel zone
TNRL	number of receiver lines	51.0	try	10.000 km for source line length
TNSL	number of source lines	27.7	try	8.100 km for receiver line length
TRL	total length—receiver lines	408.0	km	
TSL	total length—source lines	276.7	km	
	number of receivers per receiver line	161.0		
NREC	number of receiver points	8211		
NSHOTS	number of source points	5561		
NR	number of receiver stations per square km	100	/km^2	
NS	number of source points per square km	67	/km^2	
IPATCH	in-line patch size	3000	m	
XPATCH	cross-line patch size including swath	2800	m	
	cross-line patch excluding swath	2400	m	
A	aspect ratio including swath	0.93		
	aspect ratio excluding swath	0.80		
IOFFSET	in-line offset	1475	m	
XOFFSET	cross-line offset	800	m to	1400 m
X_{min} straight	largest min offset	361	m	
X_{min} straight	largest min offset—one bin offset	326	m	
X_{max}	largest max offset recorded	2052	m	***X_{max} is larger than the necessary
X_{max}				
DAYS	acquisition days	22.2	days	

Project: SEG 3-D **Client:** ABC Oil Company
Location: TEXAS **GEDCO file #:**
9/1/99

COSTS

					min	average	max	
government approvals	$2,000			$2,000				
permit agent	$500	/day		$22,000				
permitting	$0	/km^2		$0	$4,000	$7,000	$10,000	/section
receiver line permits	$400	/km		$163,200	$300	$400	$500	/mi
source line permits	$800	/km		$221,333	$800	$1,200	$1,600	/mi
general damages	$100	/km		$68,467	$100	$200	$300	/km
crop damages	$750	/ha		$79,786	$200	$300	$3,000	/acre
timber salvage	$150	/ha		$3,617	$40	$150	$154	/ha
scouting	$550	/day		$12,100	$525	$588	$650	/day
advance man	$550	/day		$24,200	$525	$588	$650	/day
cat push	$550	/day		$24,200	$500	$575	$650	/day
survey crew-perimeter	$900	/day		$19,800	$875	$913	$950	/day
bird dog	$550	/day		$12,110	$500	$575	$650	/day
health and safety management	$550	/day		$12,110	$500	$575	$650	/day
line cutting-open existing	$600	/km	171.2	$102,700	$300	$650	$1,000	/km
line cutting-cat	$1,000	/km	68.5	$68,467	$500	$1,000	$1,500	/km
line cutting-hand	$1,500	/km	239.6	$359,450	$500	$1,750	$3,000	/km
drilling	$0.30	/ft		$246,241	$0.45	$0.85	$1.25	/ft
mob/demob	$0			$0	$2,000	$3,000	$4,000	each
acquisition	$0	/shot		$0	$145	$273	$400	/shot
acquisition	$15,000	/km^2		$1,200,000	$8,000	$11,500	$15,000	/km^2
test holes	$100	for	8	$800	$80	$140	$200	/hole
clean-up	$150	/km		$102,700	$0	$350	$700	/km
survey calculations	$21	/km		$23,134	$21	$21	$21	/km
processing	$20	/shot		$111,220	$15	$19	$23	/shot
3-D design charge, min.	$925	/survey		$925				/survey
project management,								
min. estimate	$600	/day		$26,693				/day
Flexi-Bin license fee	$300	/km^2						/km^2
interpretation	$700	/day		$93,425				/day
workstation rental	$840	/day		$74,740				/day
reproduction	$700			$700	$400	$700	$1,000	
contingency	10%			$307,610				
TOTAL COST				$3,383,706				

		net of halo
cost per section		$109,541/mi^2	$150,096/mi^2
cost per km^2		$42,296/km^2	$57,955/km^2
cost per recorded trace		$0.85

Please note that all numbers are estimates only!!

prepared by: Andreas Cordsen, P. Geoph.
Geophysical Exploration & Development Corporation
#1200, 815-8th Avenue SW
Calgary, Alberta, Canada T2P 3P2
(403) 262-5780   fax (403) 262-8632
email: acordsen@gedco.com   last update: 22-Apr-99   version 3.0

4.7 Spreadsheet for 3-D Design Flowchart

Project: SEG 3-D   FOR BID REQUEST
Location: TEXAS   GEDCO file #: 1999
          9/1/99

PROSPECT INFORMATION

Units	M		M (metric) or I (imperial)

INPUT

shape	rectangle	
receiver	EW	
source line direction	NS	
receiver interval	50	m
source interval	50	m
processing bin in receiver direction	25	m
processing bin in source direction	25	m
receiver line interval	200	m
source line interval	300	m
number of receiver points	8211	
number of source points	5561	
length of receiver lines	8.000	km
length of source lines	10.00	km
number of channels	720	
patch	12 × 60	
receiver array	9	over 5 m circle
source array	1	over 50 m
roll on/off stations	yes	
swath width (max = NRL/2)	3	
number of consecutive shots which may be dropped	5	out of 80 for access reasons only—not on adjacent source lines
min. distance of any shot from edge of patch	800	m except along the edges of the survey
% of lines not needing line cuts	30	% ⎫ should total 100%
% of existing cut lines	25	% ⎬ should total 100%
% of cat cut lines	10	% ⎨ should total 100%
% of hand-cut lines	35	% ⎭ should total 100%
$ of line km on farm land with crop	20	%
% of acceptable dead traces	2	%

dynamite

number of holes	3	
hole depth	15	m
charge size per hold	1	kg

vibroseis

number of vibrators	4	
number of sweeps	8	
sweep length	12	s
pad time	96	s
sweep frequency -start	10	Hz
sweep frequency -end	90	Hz
sweep rate	6.67	Hz/s
sweep type	3	dB/octave
source interval	50	m
max move-up for each sweep	6.25	m

4.8 COST MODEL

The equation,

$$SD \times NC = \frac{Fold}{B^2}, \quad (4.12)$$

is the essence of a cost model developed by Caltex Pacific Indonesia (CPI). This model normalizes the cost of 3-D surveys with the number of recorded midpoints per unit area (Bee et al., 1994). Acquisition costs appear to have a direct relationship with data density, and one can easily determine a normalized value for data density using

3-D Data Density =

$$\frac{Fold \times U}{B^2} \quad or \quad \frac{Fold \times U}{B_s \times B_r}. \quad (4.13)$$

Example:

$$Data\ Density = \frac{25 \times 10^6}{25^2} = 40\,000\ midpoints/km^2,$$

or

$$\frac{25 \times 27.88 \times 10^6}{82.5^2} = 102\,400\ midpoints/mi^2,$$

and

$$Cost\ per\ midpoint = \frac{\$20\,000}{40\,000} = \$0.50/midpoint.$$

If the survey cost is $20 000/km² *($51 200/mi²)* for full-fold coverage, the cost per midpoint is $0.50. Making comparative calculations for the cost per midpoint of 2-D data may convince management of the cost advantage of a 3-D survey. Assuming a group interval of 20 m *(55 ft)*, 30 fold, and a cost of $6000/km *($11 520/mi)* for 2-D data, then

$$2\text{-D Data Density} = \frac{Fold \times U}{CDP\ Spacing}$$

$$= \frac{30 \times 1000}{10} = 3000\ midpoints/km, \quad (4.14)$$

or

$$= 5760\ midpoints/mi$$

and

$$Cost\ per\ midpoint = \frac{\$6000}{\$3000} = \$2/midpoint\ for\ 2\text{-}D,$$

or

$$\frac{\$11\,520}{5\,760\ midpoints} = \$2/midpoint\ for\ 2\text{-}D.$$

The typical 2-D comparative cost is $2/midpoint in the above example, which is four times the 3-D cost of $0.50/midpoint. Although the 3-D cost is lower on a per-midpoint basis, one must ask whether the additional cost of 3-D coverage is warranted or whether a higher resolution 2-D data set would suffice.

5

Field Layouts

Numerous layout strategies have been developed for 3-D surveys. One has to establish which features are important in the area of the survey in order to select the best design option. A start at such evaluation is possible with the series of displays that have been prepared for each design strategy in this chapter. All designs in this chapter are based on a recording patch of 12 lines with 60 stations per line. The aspect ratio is generally close to 0.8. Each design will be described using the following sequence of displays:

a) Layout—full scale,
b) Layout—zoomed,
c) X_{min} distribution,
d) Offset distribution within each bin,
e) Offset distribution in parallel rows of bins (within a box),
f) Azimuth distribution within each bin.

For improved comparisons, all X_{min} distributions are scaled similarly. Line spacings are comparable as well, with any exceptions noted.

5.1 FULL-FOLD 3-D

A full-fold 3-D survey is one where source points and receiver stations are distributed on an even two-dimensional grid (Figures 5.1a, 5.1b) with station spacings equal to line spacings. The grids are offset by one bin size. A full-fold 3-D survey has outstanding offset and azimuth distributions as illustrated in Figures 5.1c–5.1f, as long as one can afford to record with a large number of channels. All other 3-D designs are basically subsets of such full-fold surveys, and the designer has to decide which aspects of a 3-D design are absolutely necessary and which can be compromised.

5.2 SAMPLING THE 5-D PRESTACK WAVEFIELD

The ideal survey samples the 5-D prestack wavefield $W(t, x_s, y_s, x_r, y_r)$ that is dependent upon traveltime t and the source and receiver locations, i.e., x_s, y_s, x_r, y_r (Vermeer, 1998a.) Three-dimensional surveys generally record only a portion of this 5-D wavefield with line spacings that are greater than the station spacings. From the under-sampled 5-D wavefield that has been acquired, one can form many different 3-D subsets such as common source and receiver gathers, cross-spreads, etc. Some of the different 3-D subsets can be visualized from Figure 5.2. The trace at midpoint M is a member of the common-source gather (horizontal line), the common-receiver gather (vertical line), the common-azimuth gather (diagonal line), and the common-offset gather (circle); after Vermeer, 1998a.

Vermeer (1998a) discussed the concept of "symmetric sampling." The basic idea of this theory can best be explained by the statement, "Make common-receiver gathers look like common-shot gathers." This philosophy leads to the criteria:

shot interval = receiver interval ($SI = RI$),
shot line interval = receiver line interval ($SLI = RLI$),
maximum in-line offset = maximum cross-line offset,
center-spread acquisition for shots and receivers, and
shot arrays required as much as receiver arrays.

Application of these criteria to orthogonal geometry leads to square cross-spreads. Note that a conventional orthogonal geometry acquired with a square receiver patch can be decomposed into a series of overlapping square cross-spreads.

78   *Field Layouts*

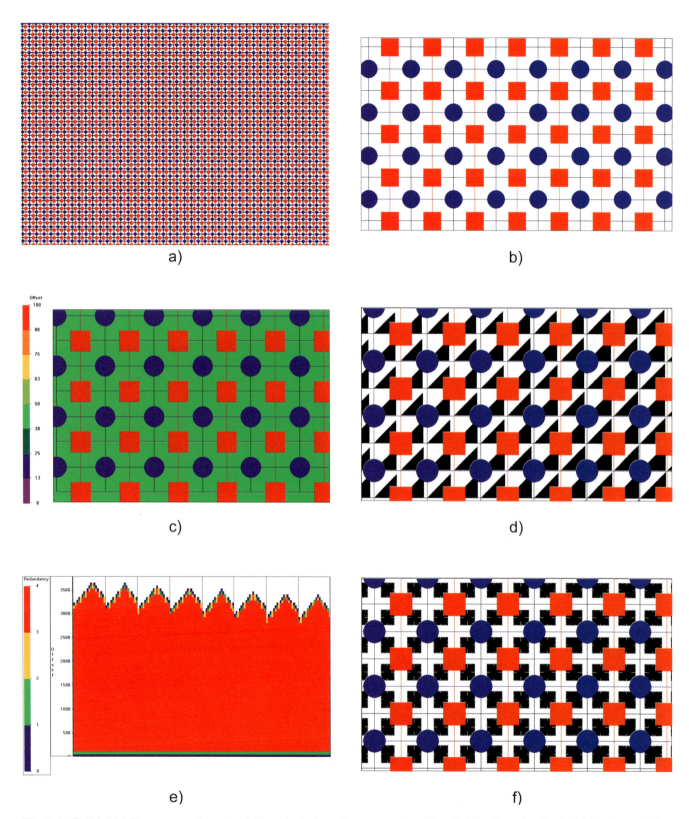

Fig. 5.1. Full-fold 3-D survey; a. layout—full scale, b. layout—zoomed, c. X_{min} distribution, d. offset distribution within each bin, e. offset distribution in parallel rows of bins (within a box), f. azimuth distribution within each bin.

5.2 Sampling the 5-D Prestack Wavefield

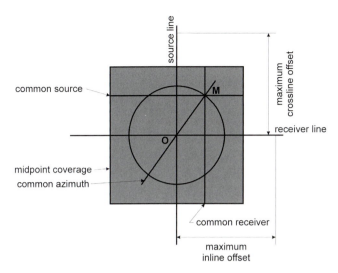

Fig. 5.2. Properties of the cross-spread (after Vermeer, 1998a). The half-offset of a trace at point M equals the distance to the center O of the cross-spread. The trace at M is part of a common-source, common-receiver, common-offset, and common-azimuth gather.

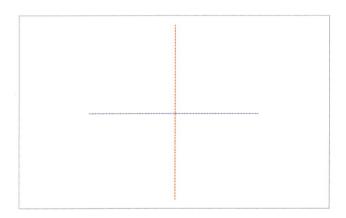

Fig. 5.3a. Cross-spread.

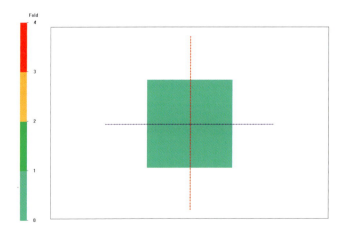

Fig. 5.3b. Cross-spread with midpoint coverage.

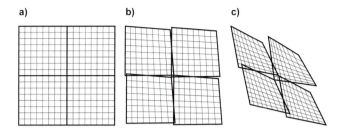

Fig. 5.4. Illumination of (a) 0°, (b) 15°, and (c) 45° dipping events by four adjacent cross-spreads (after Vermeer, 1998a).

A crossed spread (or cross-spread) has a few long source and receiver lines (generally only two lines crossing orthogonally, Figure 5.3a). This cross-spread geometry produces an area of single fold coverage (Figure 5.3b) if all receivers are recording while the source points are taken. A 3-D cross-spread could be defined as $W(t, X, y_s, x_r, Y)$, where X denotes the fact that the x-coordinate of the source does not vary, and Y indicates that the y-coordinate of the receiver is constant. If an x- and a y-coordinate are varying, a single fold continuous 3-D subset that is suitable for prestack migration is formed. The area of midpoint coverage equals half the receiver line length times half the source line length. There is a large variation in the azimuths from bin to bin. Such a single-fold data set, which can be processed to effectively image a reflector through the use of DMO or prestack migration has been defined as a minimal data set (Padhi and Holley, 1997). There are many single-fold 2-D minimal data sets that can be formed by combinations of the two source and two receiver coordinates. Proper reflector imaging can be successful as long as well-sampled minimal data sets are combined in the processing.

Spatial continuity is the key to combining minimal data sets. Even though each minimal data set is well sampled and of maximum extent, each cross-spread boundary still represents a spatial discontinuity in the 3-D data set. It is, therefore, important to maximize the useful extent of each cross-spread to minimize the overall spatial discontinuity. Spatial continuity is best served by using symmetric sampling.

The super-positioning of the overlapping midpoint areas of the cross-spreads creates the multifold coverage. In flat layered geology these midpoint areas (from square cross-spreads) overlap with a high degree of spatial continuity, which is reduced when dips are present (Vermeer, 1998a). Hence it is important to consider the effects of the possible gaps near

80　*Field Layouts*

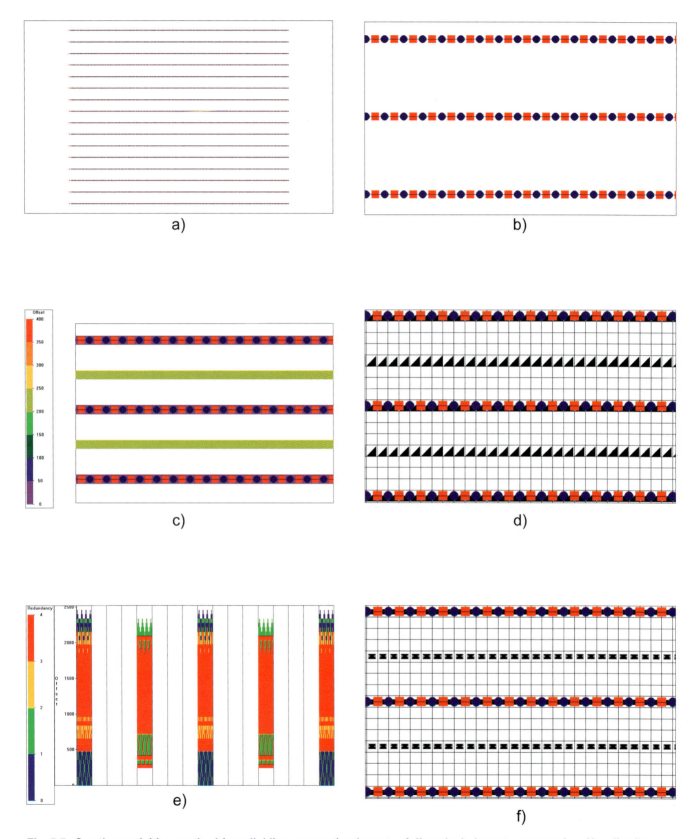

Fig. 5.5. Swath acquisition method (parallel line concept); a layout—full scale, b. layout—zoomed, c. X_{min} distribution, d. offset distribution within each bin, e. offset distribution in parallel rows of bins (within a box), f. azimuth distribution within each bin.

the edges of the midpoint areas (Figure 5.4). These effects are reduced by increasing the fold and by keeping the amount of overlap as large as possible (see Section 9.5). If, for example, the nominal fold is 30 for a 3-D survey where severe dips are present, then the areas of the gaps in the single-fold cross-spread coverage will still have 28 fold where two of such gaps are crossing each other. A 10% fold reduction might not be noticeable in this case; however, in a lower fold survey the reduction percentage is far greater and therefore the effect would be more severe.

5.3 SWATH

The swath acquisition method was used in the earliest 3-D designs (Figures 5.5a, 5.5b). In this geometry, source and receiver lines are parallel and usually coincident. While source points are taken on one line, receivers are recording not only along the source line but also along neighboring parallel receiver lines, creating swath lines halfway between pairs of source and receiver lines. The X_{min} distribution has almost zero values on the source lines and values equal to the line spacings on the swath lines (Figure 5.5c). The offset distribution in all occupied bin lines is excellent (Figures 5.5d, 5.5e). However, inadequate sampling in the cross-line direction makes this design a "poor man's 3-D," because many bins are empty. The azimuth mix is very narrow (for few receiver lines) and depends on the number of live receiver lines in the recording patch and the line spacing (Figure 5.5f). Most companies prefer to have the source points at the half-integer positions. Parallel swaths are sometimes considered on land when severe surface restrictions exist, or when costs have to be minimized. The operational advantages are attractive, but are achieved at the cost of a poor azimuth mix and poor cross-line sampling.

A gapped in-line technique has been developed to further reduce costs in the jungle environment of Indonesia (Bee et al., 1994). Gapped in-line swath shooting does not necessarily use all of the lines of a regular swath geometry as source and/or receiver lines. Swath shoots are also an inherent part of marine 3-D seismic surveys because of the practical aspect of towing air gun arrays and hydrophone cables behind a vessel. In the marine case, however, there are enough source lines and receiver lines (streamers) to create contiguous lines of midpoints (complete coverage). All such designs can be classified as parallel geometries that attempt to create continuous common-offset gathers as a 3-D subset.

5.4 ORTHOGONAL

Generally, source and receiver lines are laid out orthogonal to each other. Because the receivers cover a large area, this method is sometimes referred to as the patch method (Figures 5.6a, 5.6b). This geometry is particularly easy for the survey crew and recording crew, and keeping track of station numbering is straightforward. In an orthogonal design, the active receiver lines form a rectangular patch surrounding each source point location; creating a series of cross-spreads that overlap each other. The patch often has a longer axis in the in-line direction. The in-line offsets are usually close to the desired offsets that will be included in the stack. If the cross-line offset within the patch is close to the maximum offset for stack, then most of the receiver line farthest from the source point will be useless because much of the data recorded on that line will be muted in processing. Depending on the receiver line spacing, the aspect ratio of the axes of the patch is usually between 0.6 and 1.0. An aspect ratio near 0.85 (see The 85% Rule, Section 3.4) is usually acceptable. The source points are assumed to be located at the center of the patch, although this is not a necessity. When shooting in areas of steep regional dip, one may want to consider asymmetric patches. Alternating symmetric and then asymmetric patches are also useful for operational reasons where the survey is at the limit of the available equipment for the crew. This technique allows more surface area to be acquired prior to receiver station moves.

In the orthogonal method example, receiver lines are oriented east-west and source lines north-south (as shown in Figures 5.6a, 5.6b). This method is easy to lay out in the field and can accommodate extra equipment (layout ahead of shooting) and roll-along operation. Usually, all the source points between adjacent receiver lines are recorded, then the receiver patch is rolled over one (or more) line(s) and the process is repeated. Figure 5.6c shows the X_{min} distribution for the orthogonal design. The X_{min} for a particular bin is smallest at the line intersections and increases toward the center of the boxes. The offset distribution is good (depending on the number of channels in the recording patch), but deteriorates towards the center of the boxes where the shorter offsets are missing (Figure 5.6d). The offsets in parallel rows of bins show a tendency toward distinct patterning for offset limited stacks (tighter mutes for the shallower section), which may result in severe acquisition footprints at shallow levels (Figure 5.6e). The azimuth distribution for the orthogonal method is uniform as long as a wide recording patch is used (Figure 5.6f).

82   *Field Layouts*

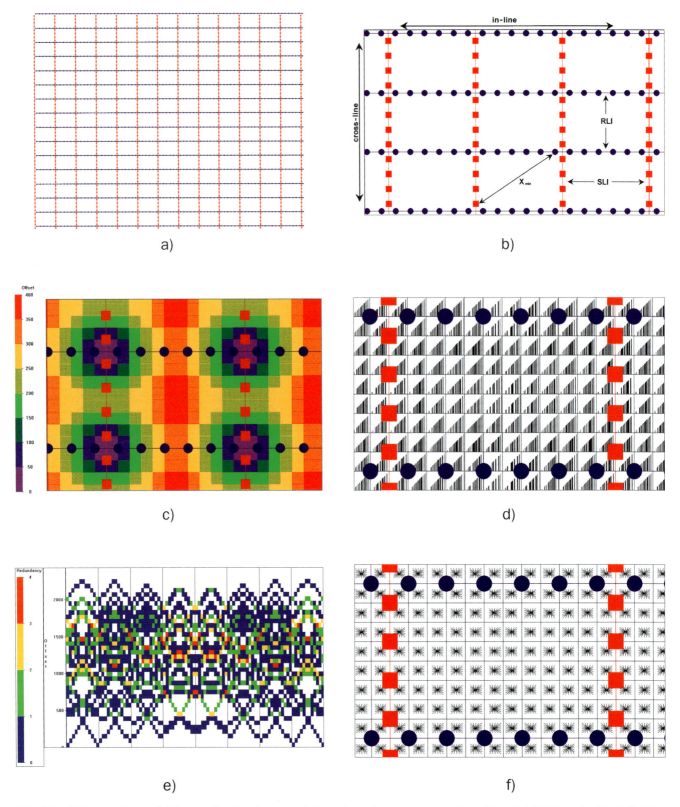

Fig. 5.6. Orthogonal acquisition method; a. layout—full scale, b. layout—zoomed, c. X_{min} distribution, d. offset distribution within each bin, e. offset distribution in parallel rows of bins (within a box), f. azimuth distribution within each bin.

5.5 BRICK

The brick pattern (Figures 5.7a, 5.7b) was developed in an attempt to improve the offset distribution of the orthogonal method. By moving the groups of source points that lie between alternate receiver lines to a half-line position, the pattern of offset distribution becomes more random in nature. In practice, it is interesting to note that a square patch (or a large aspect ratio) gives rise to more or less the same offset distribution for either the orthogonal (Figure 5.6d) or the brick method (Figure 5.7d). For a narrow azimuth patch, the offset distribution of a brick pattern is superior to that of an orthogonal design.

If one considers a typical box as defined earlier, then it can be shown that the largest minimum offset X_{min} is significantly less than in the orthogonal design assuming the same source and receiver line intervals (Figure 5.7c). X_{min} in the brick pattern design depends upon the line spacings. Because of the staggered source lines, there is no easy formula for the value of X_{min} or its location; therefore, its distribution should be confirmed with a modeling package. However, the brick design method allows one to increase the receiver line interval over the orthogonal design without increasing X_{min} as much as in the orthogonal design. If the source points and receivers are offset at the line intersections, X_{min} is slightly smaller (approximately $\frac{1}{2}$ bin) than without the offset.

The offsets in parallel rows of bins show a slight improvement over the orthogonal design; this may result in less acquisition footprint at shallow levels (Figure 5.7e). The brick design generally offers better azimuth distribution in addition to an improved offset distribution for rectangular patches (Figure 5.7f). Brick patterns are used in areas where permit costs are not an issue and easy access is provided to all locations, such as in the desert. Statics coupling remains a problem for any regular 3-D geometry except for the full-fold 3-D because the sources and receivers are uncoupled. The brick design has discontinuous common-receiver gathers because the reflectors are broken due to the bricking of the source points (Vermeer, 1998a). This leads to acquisition footprints where linear shot noise is present.

A double brick design refers to a four-line swath (Figure 5.7g). Triple brick refers to a six-line swath (Figure 5.7h), and quadruple brick to an eight-line swath (Figure 5.7i). In all of these designs, only the center line of source points is being recorded. These relatively narrow patches offer an improved offset distribution over the orthogonal design and even over a square patch of a brick pattern. It follows that one can carry this design concept to a point where all source point locations fall onto a diagonal rather than on staggered lines (see Nonorthogonal Geometries, Section 5.6).

Because of the shifting of the source lines in brick designs, it is not necessary to have in-line fold and cross-line fold independently as integers (calculated as per Chapter 2).

The traditional technique of bricking only the sources is widely accepted although one can brick the receivers instead. In areas of dense cultural interference, sources can be placed upon the roads and bricked receivers can be used to fill in the brick pattern. The technique can also be extended to a brick-brick design, where both the source and receivers are bricked with the advantage of reducing the amount of line cutting or crop damages per acre of coverage when compared to a conventional brick design.

5.6 NONORTHOGONAL

Nonorthogonal (or slanted) arrangements of source and receiver lines (Figures 5.8a, 5.8b) are used to get the benefits of the offset distributions of the brick design without some of the disadvantages, such as 90° turns and noncontinuous source lines (and therefore common receiver-gathers). For nonorthogonal designs, one needs to be careful when deciding whether to measure the station interval inline or to stretch it to fall on grid points.

Assuming a station interval of 60 m *(220 ft)*, the source points in this example should be stretched out by a factor of

$$\frac{60 \text{ m}}{\cos 45°} = \frac{60 \text{ m}}{0.707} = 85 \; m$$

or

$$\frac{220 \text{ ft}}{\cos 45°} = \frac{220 \text{ ft}}{0.707} = 311 \text{ ft}.$$

With this stretch, the bin size of 30 m × 30 m *(110 ft × 110 ft)* can be maintained, and the midpoints fall into the center of the bins. Figure 5.8a shows an example of the nonorthogonal method with a 45° angle between the source lines and the orthogonal to the receiver lines. The source points for each patch lie in the center of the patch between two adjacent receiver lines (Figure 5.8b). This basic unit is moved around the survey. This method is often shot with only a few receiver lines per patch (narrow azimuth) and is oper-

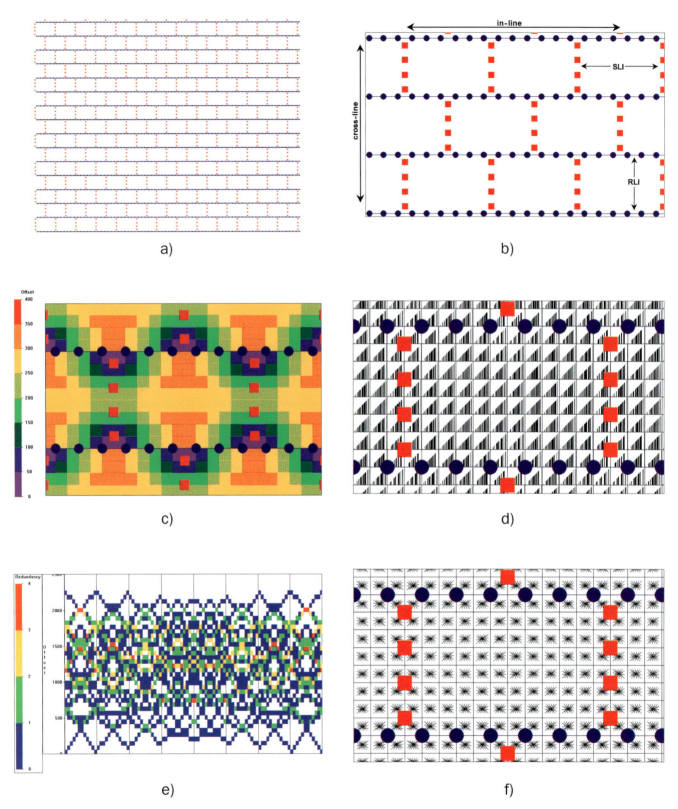

Fig. 5.7. Brick acquisition method; a. layout—full scale, b. layout—zoomed, c. X_{min} distribution, d. offset distribution within each bin, e. offset distribution in parallel rows of bins (within a box), f. azimuth distribution within each bin.

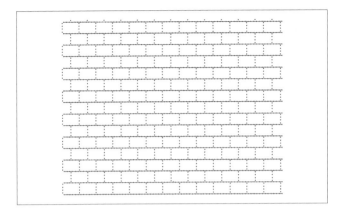

g)

h)

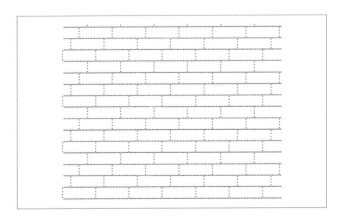

i)

Fig. 5.7. (continued) Brick acquisition method; g. double brick layout, h. triple brick layout, i. quadruple brick layout.

ationally attractive because of the straight source and receiver line layout. With a narrow azimuth patch, fold builds up quickly around the survey edges in the cross-line direction. Figure 5.8c shows the X_{min} distribution for one example of the nonorthogonal design. The position of the largest minimum offset (X_{min}) changes with the line angle and the line intervals. One should check the X_{min} distribution with a software package. The offset distribution for the 45° angle design is well dispersed with few duplicated offsets (Figures 5.8d and 5.8e). The azimuth distribution is very good also; however, it depends on the number of receiver lines in the patch (Figure 5.8f).

A special case for the nonorthogonal design is the case where the relative geometry in the cross-line direction is repeated every second receiver line (Figure 5.9a). For this case, the angle between the source lines and the orthogonal to the receiver lines is 26.565° (arc tan 0.5) when the line intervals are equal. The source points in this example should be stretched out by a factor of 60 m ÷ cos 26.565° = 60 m ÷ 0.894 = 72.1 m *(or 220 ft ÷ cos 26.565° = 220 ft ÷ 0.894 = 246 ft)*. The selection of this angle causes the midpoints to be distributed as indicated in Figure 5.9b. This midpoint distribution is not good for square binning (compare Figure 5.8b); however, it is ideal for hexagonal binning (see Section 5.11). The X_{min} distribution appears far worse than in the previous example of the 45° angle because the line spacings are greater. Finally the offset and azimuth distributions are improved over the 45° angle geometry (hence less footprint) (Figures 5.9d-5.9f).

5.7 FLEXI-BIN® OR BIN FRACTIONATION

The Flexi-Bin® method was developed and patented by GEDCO (Cordsen, 1993a, 1993b, 1995a). In this method, source points and receivers can be laid out in many different ways. Basically, one must ensure that source and receiver line spacings are noninteger with respect to the group interval. In Figure 5.10 for example, $RI = SI = 60$ m, $RLI = 260$ m, and $SLI = 320$ m. This causes midpoints to fall in a regular pattern inside each CMP bin (Figure 5.10b). In many other methods, the theoretical midpoints are at the bin center.

In the above example, (60 m, 60 m, 260 m, 320 m) nine groups of midpoints of 10 m × 10 m size result in each CMP bin of 30 m × 30 m. In processing, one can choose to stack these sub-bins or microbins (nine times as many as regular bins), thereby achieving finer midpoint (bin) spacings. With a stack trace every 10 m instead of every 30 m, these traces have

86   *Field Layouts*

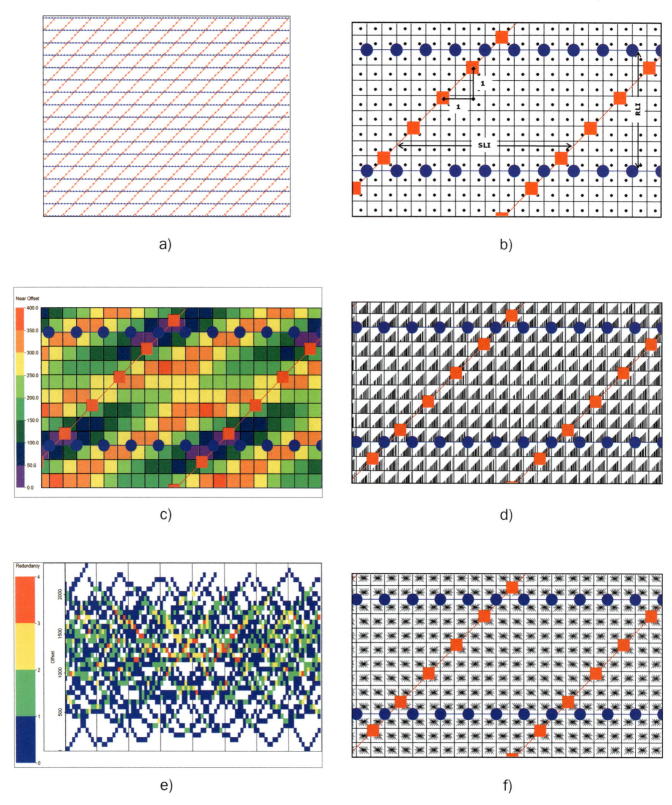

Fig. 5.8. Nonorthogonal—45°—arrangement of source and receiver lines; a. layout—full scale, b. layout—zoomed, c. X_{min} distribution, d. offset distribution within each bin, e. offset distribution in parallel rows of bins (within a box), f. azimuth distribution within each bin.

5.7 Flex-Bin© or Bin Fractionation   87

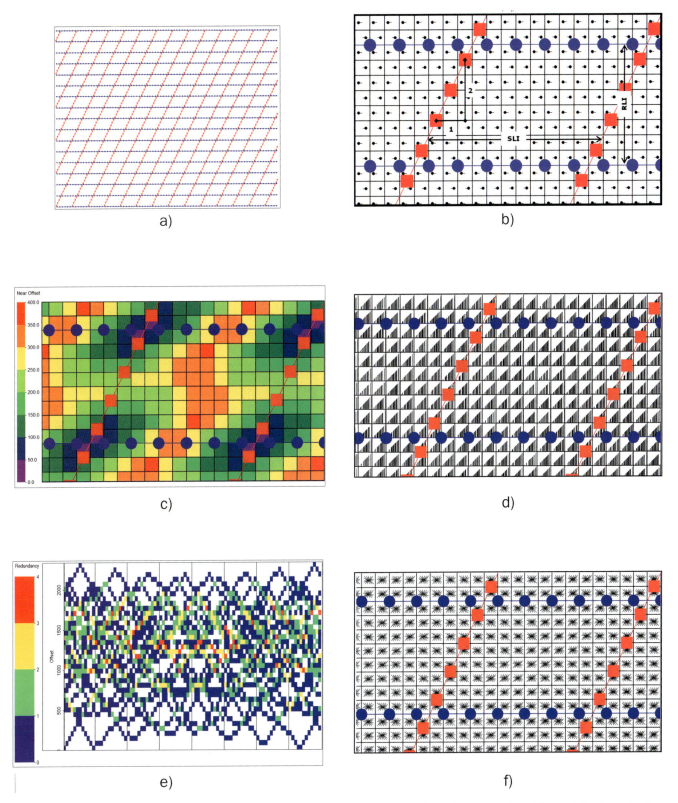

Fig. 5.9. Nonorthogonal—26.565°—arrangement; a. layout—full scale, b. layout—zoomed, c. X_{min} distribution, d. offset distribution within each bin, e. offset distribution in parallel rows of bins (within a box), f. azimuth distribution within each bin.

88   Field Layouts

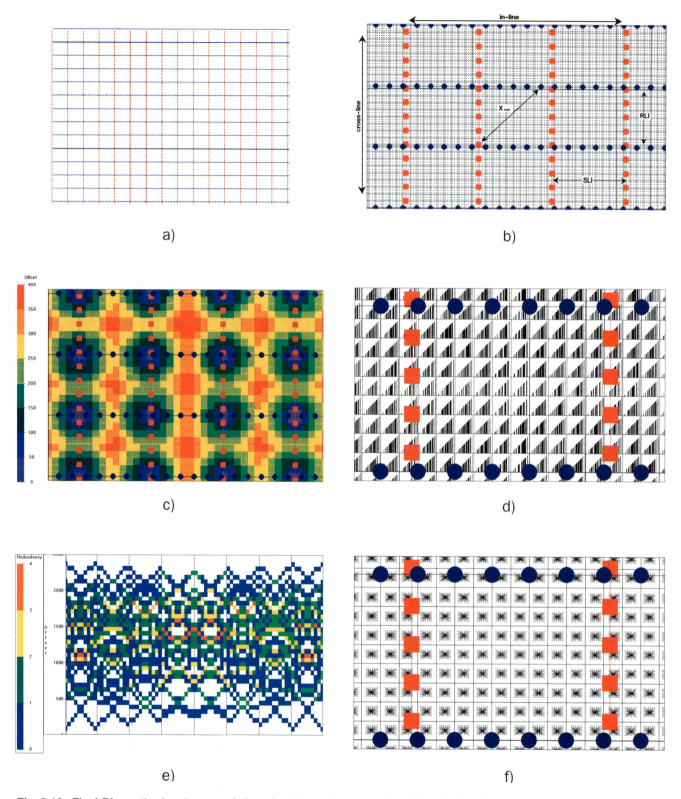

Fig. 5.10. Flexi-Bin method; a. layout—full scale, b. layout—zoomed, c. X_{min} distribution, d. offset distribution within each bin, e. offset distribution in parallel rows of bins (within a box), f. azimuth distribution within each bin.

lower fold, and therefore, potentially lower signal-to-noise ratio. On the other hand, one can improve the S/N by increasing the bin size. This decision can be made on the basis of data quality after the data are recorded.

The X_{min} distribution is similar to the orthogonal layout (Figure 5.10c). Each normal CMP bin still contains as many traces as before, but with a somewhat improved offset (Figures 5.10d, 5.10e) and azimuth distribution (Figure 5.10f). Full statics coupling (see Chapter 10) is a major advantage of this method because of the nonrepeating geometry at neighboring line intersections. The possibility exists that the smaller stack trace interval may reveal a feature that one might miss at larger stack trace spacing. There are numerous operational advantages to the Flexi-Bin® method, such as easier field operations, excellent statics coupling (Cordsen, 1995b), and other processing flexibilities. In orthogonal shooting, one should consider using this noninteger method.

The bin fractionation method (Flentge, 1996) uses source and receiver lines that are staggered by a selected fraction of the source station and receiver station intervals. GRI (1994) describes a survey where the fractionation was one-half of the source- and receiver-station intervals. That survey was limited to creating "quarter cells" in the subsurface. The numbering of station locations in the field is not parallel from line to line; however, this does not create any operational problems if reasonable care is exercised.

It has been emphasized (Vermeer, 1997) that any bin fractionation technique merely interleaves two or more cross-spreads with their sampling intervals corresponding to the station spacings. Nevertheless the subsurface samples are distributed on a finer grid of midpoints, in particular for subhorizontal events, and may indeed be beneficial for subsurface definition. Field practice with the Flexi-Bin® method on numerous seismic programs has demonstrated some improvement in lateral interpretability.

5.8 BUTTON PATCH

The button patch method was developed and patented by ARCO (Bremner et al., 1990) and has routinely been used in many ARCO 3-D surveys. Each button contains a tight pattern of receivers (Figures 5.11a, 5.11b), typically 6 × 6, 6 × 8, or 8 × 8. The final button geometry is largely determined by equipment considerations and cable restrictions. There is no requirement to keep the receiver patterns square.

Several buttons are combined in a checkerboard pattern to form the receiver patch. Multiple source points are fired into the receiver patch in a precise manner. The receiver patch is then rolled to its next location, while some overlap with the previous patch position is maintained. Then a similar pattern of source points is fired into the new receiver patch. Repeating source points for different buttons provide improved statics coupling, while staggered source points between previous locations offer better midpoint distribution. Frequently, the receiver buttons and source point locations are distributed irregularly because of surface obstructions.

Large channel capacities are necessary to minimize receiver moves. The shooting trucks, or vibrators, must travel around the patch for each new receiver layout. If sufficient receivers are available, then shrewd use of the roll-along switch can eliminate unnecessary movement of the shooters or vibrators. Two or three patch positions with roll-on can often be covered in one receiver layout.

The receiver buttons are laid out and moved over the area to be imaged at full-fold. Typically smaller bins with lower fold are used in button patch designs as compared to other design strategies. Source points are placed outside this full-fold area to give a fold taper around the edges. This geometry achieves longer offsets without the need to lay out additional equipment outside the planned area of the survey. This technique tends to improve migration and DMO because the seismic amplitudes contained on these long offsets contribute energies to many trace gathers. Because of the discontinuity of common-source gathers, this geometry may have a significant acquisition footprint.

The button patch method utilizes modern, high-channel systems effectively. High resolution can be achieved by using closer receiver spacing. A small X_{min} is present within the receiver buttons (Figure 5.11c); however, the X_{min} distribution worsens significantly outside the buttons. Good static coupling requires great care at the planning stage. Short-offset distribution may be poor, but far-offset distribution should be good because the source points are outside of the button patch area (Figure 5.11d). The offsets in a row of bins look good, but display patterning for near offsets (Figure 5.11e). Some azimuthal bunching of source-to-receiver raypaths exists because of the button layout of the receivers (Figure 5.11f).

One advantage of this design is that it has the flexibility to plan the button holes around obstacles. There is also additional flexibility for make-up source points

90   *Field Layouts*

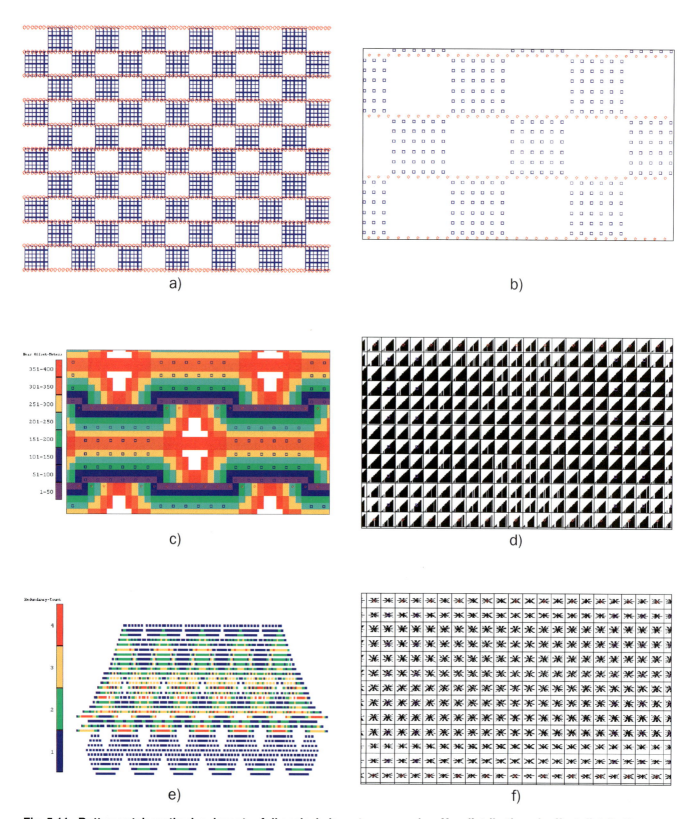

Fig. 5.11. Button patch method; a. layout—full scale, b. layout—zoomed, c. X_{min} distribution, d. offset distribution within each bin, e. offset distribution in a row of bins (along a receiver line), f. azimuth distribution within each bin.

and for compensating for dropped source points. The major constraint on the efficiency of the button patch design is the requirement for the shooters or vibrators to move about the survey easily. In some surveys, limited access may make this strategy a poor choice.

5.9 ZIG-ZAG

The zig-zag pattern (Figure 5.12a) is popular in desert areas, or other locations where one has good access between receiver lines. Single source lines are located between adjacent pairs of receiver lines for a single zig-zag (Figure 5.12b). The source point positions should be located on a grid to create central midpoints. For 60 m *(220 ft)* station intervals, and a 45° angle between the receiver lines and the source line diagonal, the distance between stations on the diagonal is 85 m *(311 ft)*. In a mirrored zig-zag, every second source line is flipped to its mirror image (Figure 5.13a, 5.13b). The offset distribution can be improved further by shooting the 3-D grid in a double zig-zag pattern, as shown in Figures 5.14a and 5.14b with two sets of vibrators.

For the zig-zag, mirrored, or double zig-zag, the largest minimum offset X_{min} is usually found near the center of the open area left by the zig-zag source lines (Figures 5.12c, 5.13c, 5.14c). It is recommended the largest X_{min} be checked using a modeling program. Zig-zag designs are often considered for narrow azimuth surveys that require good offset distribution. The offset distributions for the single (Figure 5.12d) and mirrored zig-zag designs (Figure 5.13d) are good (using a wide patch); for the double zig-zag design it is outstanding (Figure 5.14d). Figures 5.12e, 5.13e, and 5.14e indicate that there is a stronger possibility of acquisition footprint with the single zig-zag designs than with a double zig-zag. Note, however, that the source effort is increased by a factor of two for the double zig-zag. The azimuth distributions are acceptable for a wide-azimuth patch (Figures 5.12f, 5.13f, 5.14f).

The advantages of a zig-zag design are in efficiency of movement with the vibrator source along very long source lines; vibrators do not have to cross receiver lines. When compared to a brick or orthogonal design, the zig-zag has minimal moveup time between source points and effectively no drive around time. This is particularly true when two vibrator sources are used on a zig-zag design compared to a double brick. The disadvantage of the zig-zag design is that, if improperly designed, a strong acquisition footprint can be impressed upon the data.

5.10 MEGA-BIN

This term was coined by PanCanadian for a 3-D design method developed by Goodway and Ragan (1995). The geometry is based on several concepts that are brought together in a unique way. Any asymmetry between the station and line dimensions of conventional designs aliases nonrandom, source-generated surface noise. The redistribution of the source and receiver locations in the mega-bin design (Figures 5.15a, 5.15b) reduces this asymmetry and samples such noise better. In addition, the acquisition footprint, which is typical for wide line spacings, is significantly reduced.

Signal-to-noise ratio improves with fold. With mega-bin, the statistics improve for one row of high-fold bins (and hence higher S/N ratio), while bins in the neighboring row have zero fold, i.e., these bins are empty (Figure 5.15c). These latter bins are filled through premigration *f-x* interpolation from the higher fold bins. The *f-x* interpolation makes this technique different from the earlier swath surveys (see Section 5.3). The basic geometry of these two methods, however, is comparable; the line spacing in the swath survey was never set to any particular interval, while in the mega-bin technique the line spacing is generally four times the bin spacing.

The X_{min} distribution has almost zero values on the source lines and values equal to the line spacings on the swath lines (Figure 5.15c). The offset distribution on all bin lines is excellent (Figure 5.15d). No other method (other than full-fold or swath) covers all offsets in the occupied rows of bins as well as mega-bin (Figure 5.15e); however, only every other bin is occupied. The azimuth distribution is also excellent on the acquisition lines as well as on the intermediate lines (Figure 5.15f).

The data quality of mega-bin surveys has been proven to be high in relatively unstructured areas, and independently processed decimation experiments with an over-sampled 3-D survey (shots and receivers on a 70 m × 70 m grid) have shown this method to be viable. One major advantage of this method is, depending on the design, that the source points are placed along the receiver lines, thereby reducing the line cutting. However, the receiver lines are placed much closer together than in an orthogonal geometry.

5.11 HEXAGONAL BINNING

If the midpoints fall into a hexagonal pattern, the midpoints can be binned hexagonally. This method requires 13.4% fewer subsurface samples

92   *Field Layouts*

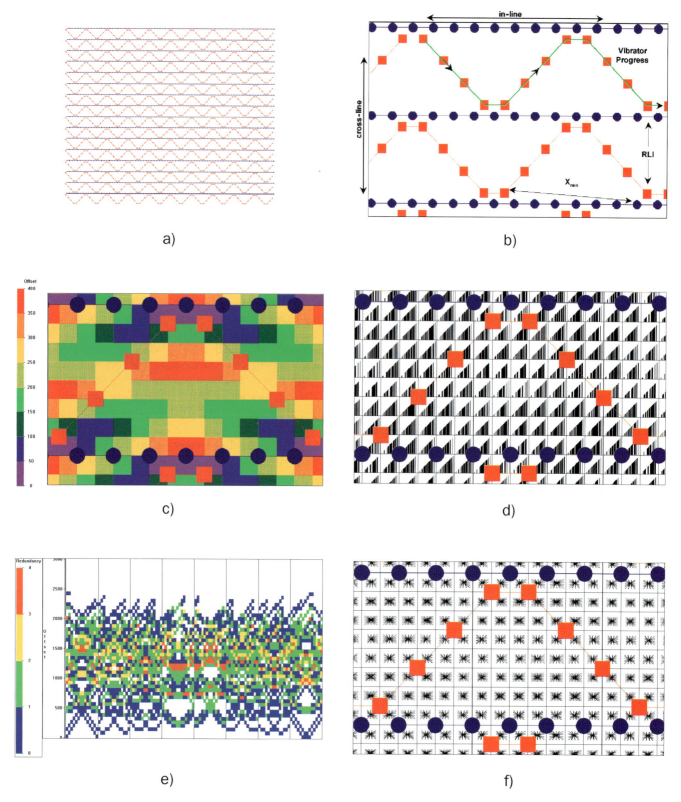

Fig. 5.12. Zig-zag method; a. layout—full scale, b. layout—zoomed, c. X_{min} distribution, d. offset distribution within each bin, e. offset distribution in parallel rows of bins (within a box), f. azimuth distribution within each bin.

5.11 Hexagonal Binning

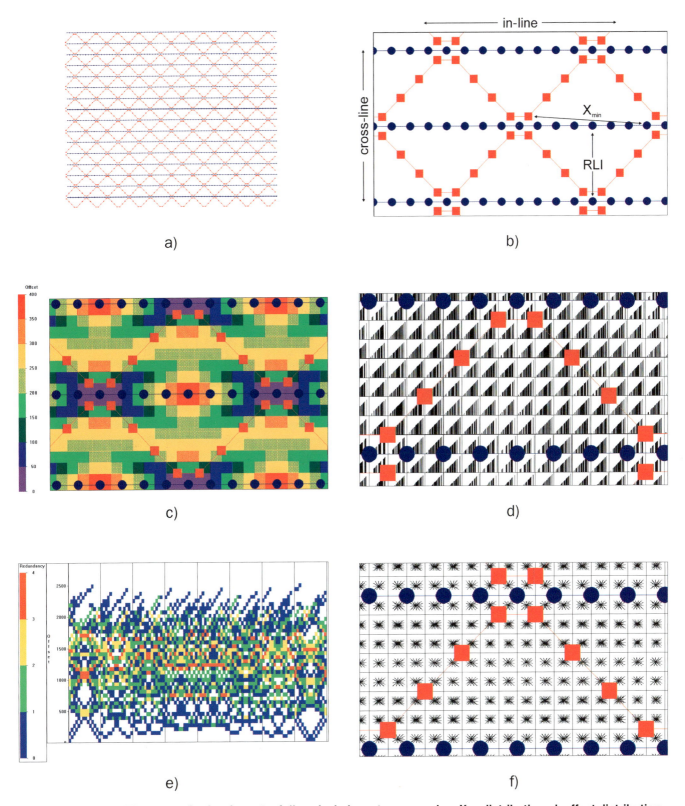

Fig. 5.13. Mirrored Zig-zag method; a. layout—full scale, b. layout—zoomed, c. X_{min} distribution, d. offset distribution within each bin, e. offset distribution in parallel rows of bins (within a box), f. azimuth distribution within each bin.

94 Field Layouts

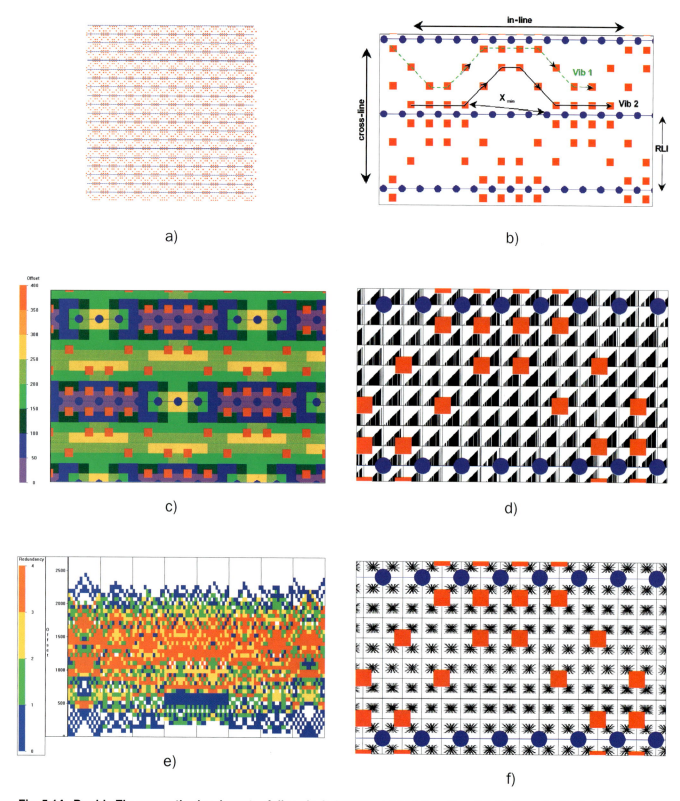

Fig. 5.14. Double Zig-zag method; a. layout—full scale, b. layout—zoomed, c. X_{min} distribution, d. offset distribution within each bin, e. offset distribution in parallel rows of bins (within a box), f. azimuth distribution within each bin.

5.11 Hexagonal Binning 95

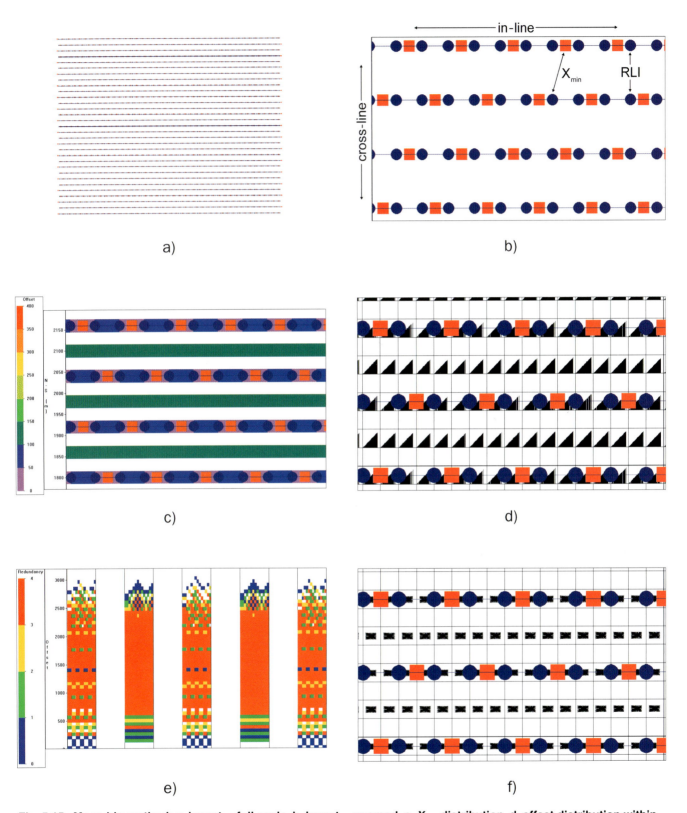

Fig. 5.15. Mega-bin method; a. layout—full scale, b. layout—zoomed, c. X_{min} distribution, d. offset distribution within each bin, e. offset distribution in parallel rows of bins (within a box), f. azimuth distribution within each bin.

in one direction compared to a square grid of midpoints, thereby reducing the acquisition and processing effort of the 3-D survey (Bardan, 1997). Either the source or the receiver effort can be reduced by 13.4% and yet still achieve the same resolution. Hexagonal sampling can be accomplished by many different layout methods, one of which was decribed earlier (see Figure 5.9b).

5.12 STAR

Star shooting involves laying out receiver lines in an arrangement resembling the spokes of a wheel (Figures 5.16a, 5.16b). Source points are taken along those same lines. In such surveys one can often have all receivers live while acquiring the source points. Fold is very high near the center and drops off quickly towards the edges of the survey. A method of improving the areal coverage is to offset the lines at their intersections. The X_{min}, offset, and azimuth distributions deteriorate very quickly away from the center (Figures 5.16c–5.16f).

The field effort for a star survey is very low compared to other designs since the star survey is essentially a series of crossing 2-D lines. Because of the high fold, many source points near the center of the survey probably can be eliminated depending on shallow horizon requirements. This method has been used with considerable success for small structures where data showed high S/N content. The method has been advocated for salt-domes (or reefs), where this geometry favors the collection of energy that reflects from all sides of the salt-dome, and long offsets are required for migration purposes. In practice, however, the location of the top of the salt-dome has to be known.

5.13 RADIAL

An improvement over the star design is the radial design. This concept involves laying out receiver lines in an arrangement resembling the spokes of a wheel similar to the star design. However, source points are placed along concentric circles around the center of the survey (Figures 5.17a, 5.17b). In such surveys one can often have all receivers live while acquiring the source points. Fold is superior to that achieved with the star design because of the different source layout but still drops off from the center toward the edges of the survey. This method enhances the areal coverage over the star design. The X_{min}, offset, and azimuth distributions also deteriorate very quickly away from the center (Figures 5.17c–5.17f).

The field effort for a radial survey is very low compared to other geometries because of the focused acquisition. Because of the high fold, many source points near the center of the survey probably can be eliminated, depending on shallow horizon requirements. Radial surveys, however, may be practical only for structures whose locations are known (see previous section). The field effort (number of source and receiver locations) for the star and radial surveys is comparable, with the exception being that the radial design requires more lines because the sources are laid out on additional concentric circles. In the case of salt dome targets, only the receivers on the same side of the target as the sources are used; this reduces imaging problems caused by complex raypaths.

5.14 RANDOM

Regularity in 3-D design often has a deteriorating effect on offset and azimuth distribution, at times causing a severe acquisition footprint (Cordsen, 1999). Smaller *RLI* to *SI* and *SLI* to *RI* ratios reduce the footprint effects in conventional designs, but cost and other considerations rarely allow such smaller line spacings. If proper sampling of the 5-D wavefield is not possible, then random acquisition can be considered to reduce the migration noise.

The trend in 3-D designs has been to irregularize the offset and azimuth distribution, but still keep field operations feasible (i.e., straight lines, short driving distances, etc.). Sometimes this irregularity has been accomplished with rather arbitrary offsets and skids. At other times surface restrictions necessitate moves of both sources and receivers. Generally, receiver moves are smaller because of takeout restrictions, and source moves are larger because of distance requirements from obstacles. The random technique has been applied successfully where source and receiver positions are moved based upon the relative ease and safety of crew during layout. This natural randomization has an added benefit of increasing production of the layout crew by minimizing the time spent attempting to locate points in regions that are difficult to access.

The main advantages of a true randomization of sources and receivers are improvements in the offset and azimuth distribution. Complete static coupling is also achieved. The theoretical fold distribution is not as smooth at the bin-to-bin level; however, the nominal fold can still be calculated according to Sections 2.3–2.7. The minimum and maximum offsets

5.14 Random   97

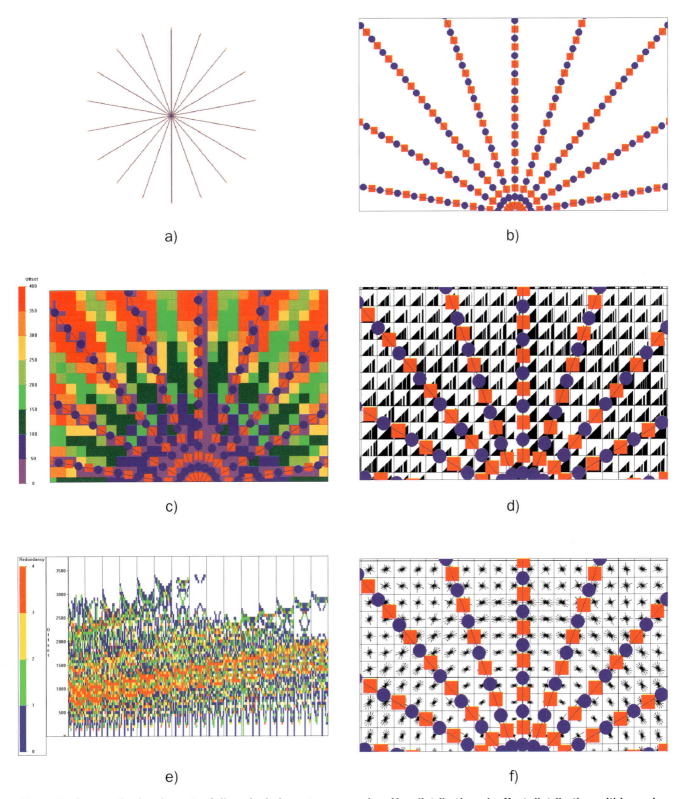

Fig. 5.16. Star method; a. layout—full scale, b. layout—zoomed, c. X_{min} distribution, d. offset distribution within each bin, e. offset distribution in several rows of bins, f. azimuth distribution within each bin.

98   Field Layouts

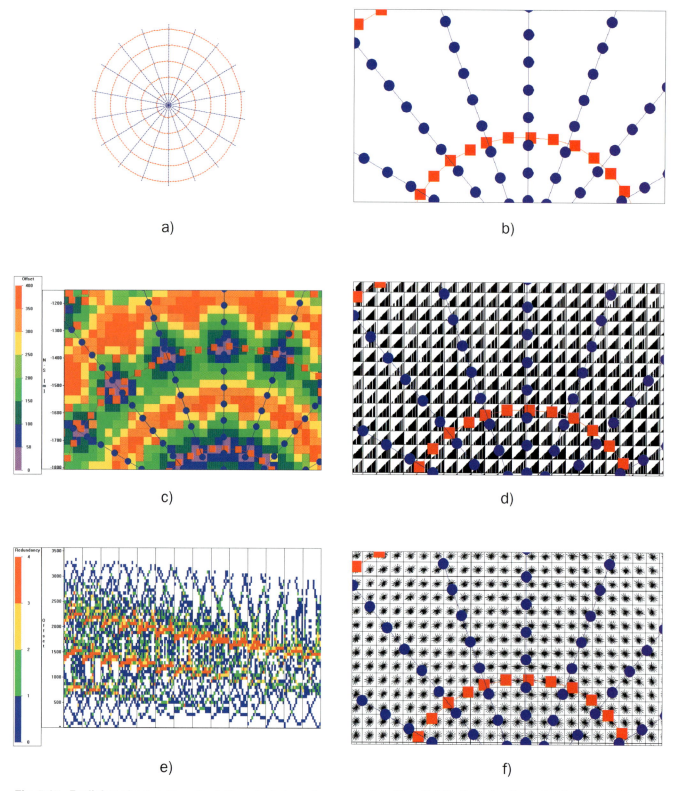

Fig. 5.17. Radial method; a. layout—full scale, b. layout—zoomed, c. X_{min} distribution, d. offset distribution within each bin, e. offset distribution in several rows of bins, f. azimuth distribution within each bin.

also vary more than in a conventional design. Both of these effects are believed to be of minor importance, because the variations within the neighborhood of bins ought to be considered (~ the size of the Fresnel zone) and within such an area the variations should be small.

Figures 5.18a and 5.18b show one possible solution for a random layout. A randomization function is applied to each source and receiver station of the nonorthogonal layout of Figure 5.9b, with the stations being allowed to move up to one bin interval in-line and two bin intervals cross-line (Cordsen, 1999). Sources are located within an area equal to $SI \times 2RI$, while receivers should be placed within an area equal to $RI \times 2SI$ centered around each original (theoretical) location. Maximum randomization is probably possible only if access is wide open, and dynamite sources and a telemetry recording system are used.

The operational and permitting advantages of the random design should also be considered as strengths of the design. Since there is not a regular pattern to the survey, there are less obvious effects on the final stack volume than in a more conventional survey because of skips caused by no permit zones or adjustments made for environmental restrictions. It becomes very difficult to tell what was an intentional source or receiver move to randomize a survey or a move caused by a safety or access concern. Finally, as far as permitting is concerned, giving the permit agents leeway to move points during initial permitting can reduce the total permit budget and cost by reducing the total effort spent occupying stations of a design preplot.

Randomization of only one set of stations (i.e., either source or receiver) introduces striping in the X_{min} distribution as well as other criteria. Best results are achieved by randomizing both source and receiver locations. In surveys which have coarse regular sampling (and therefore migration noise), Schuster and Zhou (1996) have confirmed that a quasi-random grid of sources shot into a quasi-random grid of receivers leads to a reduction of migration aliasing artifacts in the final image. Such field implementation could significantly reduce acquisition costs. Random surveys should be considered whenever coarse sampling is done because large bin sizes lead to spatial aliasing of the desired information.

The X_{min} distribution is more random than previously presented (compare Figures 5.18c and 5.9c). The offset and azimuth distributions are of course very random (Figures 5.18d–5.18f). The offsets in parallel rows of bins have no pattern at all, which is beneficial for reducing the acquisition footprint (Figure 5.18e).

5.15 CIRCULAR PATCH

The main feature of a circular patch is that all live receiver groups are limited to lie within a maximum offset (or radius) from the source point (Figure 5.19a). This method can be applied to any of the layout strategies presented in this chapter. The following figures should be compared to the orthogonal layout in Section 5.4. The X_{min} distribution is exactly the same (Figure 5.19c); the offset and azimuth distributions (Figures 5.19d–5.19f) vary only at the far offsets because of the X_{mute} limit.

The main feature of a circular patch is the maximum offset (or radius). By using computer programs to identify all stations within a particular radius, and by identifying the live stations for each receiver line, one can avoid recording stations beyond X_{mute} from the source (Figure 5.19a). The advantage of the circular patch is that the patch size can be set to be the maximum useful data size; for example, the patch size could be the mute pattern that intersects the basement reflector. Circular patches are not popular in the field because of the operational difficulties and are difficult to implement without computer modeling software and significant interaction with the crew in the field. Circular patches are used in surveys that have highly irregular receiver line spacings, such as random designs, where no conceptual patch exists.

5.16 NOMINAL FOLD COMPARISON

A comparison of the nominal fold between the various 3-D designs is useful to fully appreciate the variations in fold distribution in the subsurface (Figure 5.20). A 12 × 60 receiver patch has been maintained as much as possible in all fold calculations. This patch provides a nominal fold of 30 in each bin in the center of the survey. All designs were scaled to a maximum of 35 fold.

In the swath method (Figure 5.20a) there are two lines of fold coverage for every source line, as long as the neighboring receiver lines are recording while the center source line is shooting. This does not provide sufficient coverage for a continuous sampling of the wavefield.

The orthogonal, brick, nonorthogonal, Flexi-Bin®, button, and zig-zag designs create an even 30 fold throughout the center of the survey (Figures 5.20b–5.20h). The main differences between these designs from a fold point of view are the fold coverage along the edges of the survey. Other key features

100 *Field Layouts*

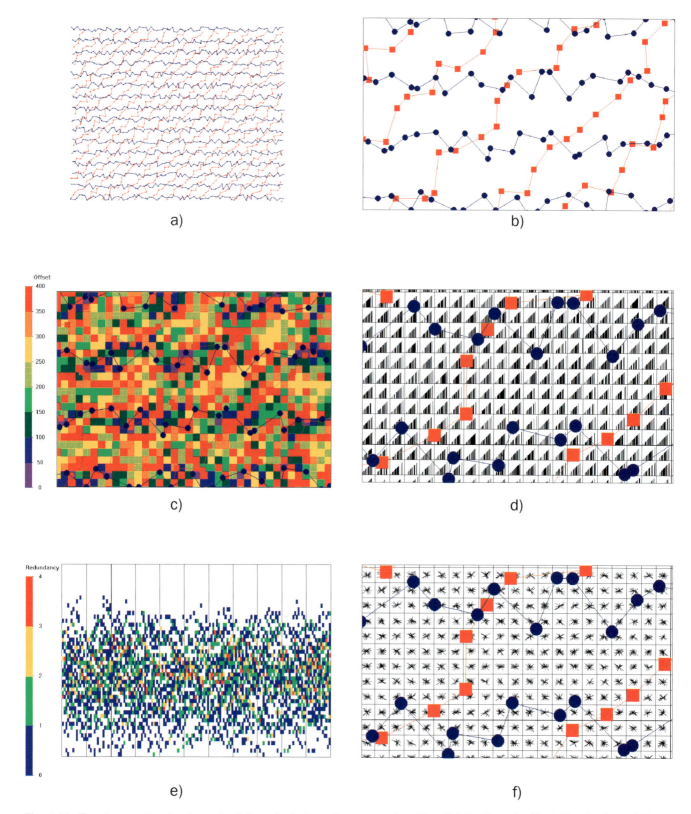

Fig. 5.18. Random method; a. layout—full scale, b. layout—zoomed, c. X_{min} distribution, d. offset distribution within each bin, e. offset distribution in several rows of bins, f. azimuth distribution within each bin.

5.16 Nominal Fold Comparison 101

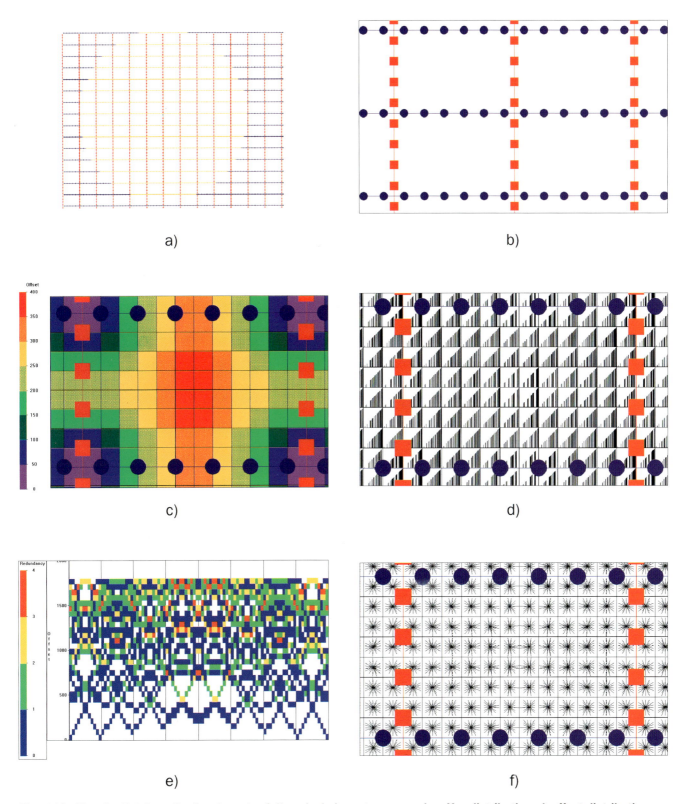

Fig. 5.19. Circular Patch method; a. layout—full scale, b. layout—zoomed, c. X_{min} distribution, d. offset distribution within each bin, e. offset distribution in parallel rows of bins (within a box), f. azimuth distribution within each bin.

102   Field Layouts

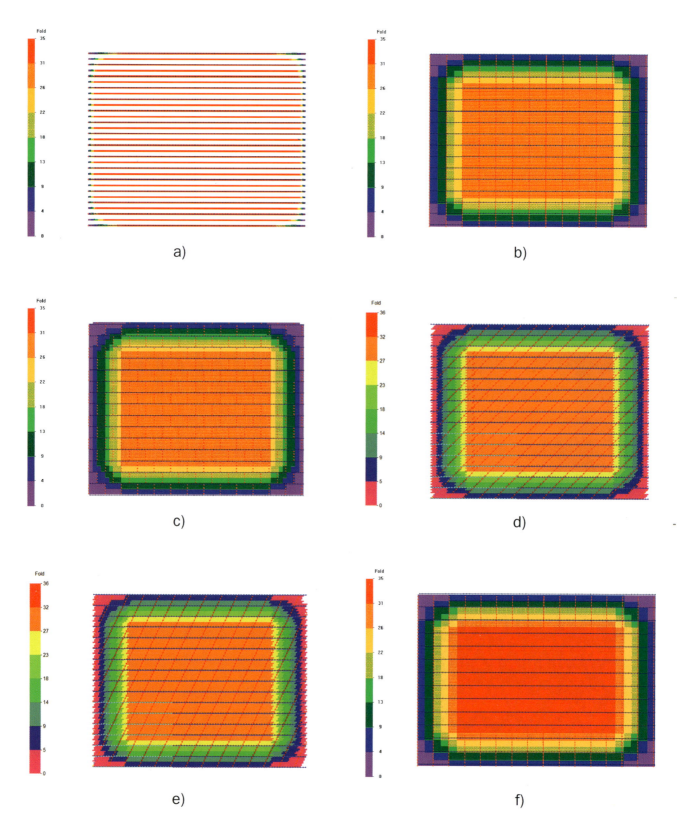

Fig. 5.20. Full-fold comparison; a. Swath, b. Orthogonal, c. Brick, d. Nonorthogonal—45°, e. Nonorthogonal—26.565°, f. Flexi-Bin.

5.16 Nominal Fold Comparison 103

Fig. 5.20. (continued) Full-fold comparison; g. Button Patch, h. Zig-zag, i. Megabin, j. Star, k. Radial, l. Random.

for these designs have been discussed earlier in this chapter.

The mega-bin design has two lines of fold coverage for every source line, similar to the swath method (Figure 5.20i). The major differences are that the recording patch now is far greater than with the earlier swath method and the line interval is four times the bin size. However, one can certainly see the similarity between these two designs.

The star and radial designs reach far greater fold in the center of the survey than the nominal 30 fold of the other designs (Figures 5.20j, 5.20k). It therefore may be desirable to offset the lines somewhat in the center or eliminate some source points near the center. The radial design offers far better fold coverage than the star design toward the edges of the survey.

A circular patch produces an even fold distribution similar to the earlier Figures 5.20b–5.20h, because it depends merely on the selection of the offset distance of receivers to be included (which should be equal to X_{mute}). Any random design creates some higher fold bins at the expense of some lower fold coverage in other bins (Figure 5.20e).

Table 5.1 compares some of the major advantages and disadvantages of each of the design methods presented in this chapter. It is by no means a complete-treatment; the reader is referred to the individual sections of this chapter for further information.

Table 5.1. Field layouts—pros and cons of various layout strategies.

Layout	Pros	Cons
Swath	Simple geometry. Cost efficient. Good offset distribution. Minimum equipment movement.	Poor azimuth distribution. Poor statics coupling.
Orthogonal	Simple geometry.	Large X_{min}.
Brick	Smaller X_{min} may allow a wider *RLI*. Reasonable offsets and azimuths.	Access can be a problem. Poor sampling in common-receiver gather can lead to acquisition footprint.
Nonorthogonal	Simple geometry.	Same as orthogonal.
Flexi-Bin® or Bin Fractionation*	High resolution with low fold, or low resolution with high fold. Super bins for normal use have good offset and azimuth mix. Excellent statics coupling.	Same as orthogonal.
Button Patch*	Efficient utilization of large channel systems. Good offset and azimuth distribution require detailed planning.	Can require large number of source points over a wide area for each patch. Needs large channel capacity. Static coupling hard to accomplish. Prone to acquisition footprint.
Zig-Zag	Same as brick. Efficient for equipment moves.	Must have very open access. Single zig-zags are prone to acquisition footprint
Mega-Bin*	Improved noise sampling. Reduced X_{min}.	Similar to swath method. Must *f-x* interpolate to fill empty bins.
Star and Radial	Good for salt domes. Offers excellent offsets for migration.	Best with all lines live. Highly irregular statistics.
Random	Improved offset and azimuth distribution. Minimal acquisition footprint.	Fold, X_{min} and X_{max} are more random.
Circular Patch	Consistent X_{max}.	Operationally difficult.

*Patent Restrictions apply to the use of these technologies.

Chapter 5 Quiz

1. Where is the bin that contains the largest minimum offset in an orthogonal design?
2. What does the largest minimum offset in a brick pattern design depend on?
3. What is the basic concept of Flexi-Bin®?
4. What is essential for the application of the zigzag method?
5. What is the difference between a star and a radial layout?

6

Source Equipment

6.1 EXPLOSIVE SOURCES

Explosive sources produce robust *P*-waves. The selection of explosives as the source of choice depends primarily on near-surface conditions and the accessibility of other energy sources. If drilling is fast and efficient, single shotholes filled with explosives might be the most economical source option. The explosive source consists of a detonator (Figure 6.1) and an explosive charge (Figures 6.2 and 6.3). In the seismic industry the explosive charge is commonly referred to as "powder" and the detonators are referred to as "caps" or "primers."

Terrain conditions may also dictate using dynamite as a source. Generally the cost of explosives may be less compared with vibrators, and the availability is better in some parts of the world. Heli-portable operations in mountainous regions seem to always require dynamite shooting. However, if a shothole pattern is required, or if the depth of the shothole pattern exceeds 10–15 m *(30–50 ft)*, the cost of explosives may greatly exceed the cost of vibrators.

Heli-portable drilling requires the use of a helicopter and lightweight drill rigs (Figure 6.4). These drill rigs have three basic parts: the frame and mast, the basket for tools and explosives (Figure 6.5), and the prime mover (Figure 6.6). The procedure for heli-portable drilling is to assemble the rig and begin drilling the hole. After the hole is drilled, the source has to be "made up" (Figure 6.7), which must be done safely. The concern here is that the primer has been mated to the "shot" or source and stray electric currents could set off the source on the surface and seriously injure anyone nearby. Once a charge is made up, it must be lowered into the hole (Figure 6.8) and the hole back-filled before detonation to prevent a blowout (Figure 6.9).

Conventional or buggy drilling is a similar procedure except that the rig drives to the site instead of being transported by a helicopter (Figure 6.10). As with heli-portable drilling, an explosive and detonator magazine is still required (Figure 6.11). All other procedures are identical to heli-portable drilling.

Surface shots are another type of energy source used when access is limited. A surface shot can be a shaped charge or a Poulter shot. A Poulter shot uses bags of explosives (Figure 6.12) that are placed above the ground on stakes to form an array (Figure 6.13) and then fired (Figure 6.14). Another type of surface shot is the shaped charge (Figure 6.15). A shaped charge is a specially designed explosive that attempts to directionally force the energy of the source into the ground. There are two other sources commonly used in surface shooting: the Primacord trench and the shallow-hole explosive (Figure 6.16). The Primacord source uses a trench gouged into the earth that has a thin tube filled with explosives placed along the length of the trench. The trench is then back-filled and detonated. The minihole source is similar to conventional deep explosive sources, except the charges are smaller, the holes are shallower (in the range of 2 to 5 m deep), and smaller, more mobile drill rigs are used (Figure 6.17).

A variety of other sources (e.g., air gun, weight-drop, p-shooter) are available when explosives or vibrators cannot be used. Unusual surface conditions or geophysical requirements are usually the driving force toward considering nonstandard sources.

With explosives, the source effort depends on the following factors:

charge size
charge depth
number of holes per source location
source type

Fig. 6.1. Detonator.

Fig. 6.2. Explosive charge.

Fig. 6.3. Explosive charges in storage boxes.

Fig. 6.4. Helicopter transport of drilling rig.

Fig. 6.5. Tool basket.

6.1 Explosive Sources 109

Fig. 6.6. Heli-portable drilling operations.

Fig. 6.8. Lowering of charge into shothole.

Fig. 6.7. "Making up" the dynamite charge.

Fig. 6.9. Backfilled shothole.

110   Source Equipment

Fig. 6.10. Conventional drilling rig.

Fig. 6.12. Poulter bag.

Fig. 6.11. Explosive and detonator magazine.

Fig. 6.13. Poulter source array.

6.1 Explosive Sources   111

Fig. 6.14. Poulter shot.

Fig. 6.15. Shaped charge source array.

Fig. 6.16. Different source types.

Fig. 6.17. Small drilling rig for minihole explosives.

The choice of charge size depends largely on the depth to the horizon of interest. The best charge size is that which achieves the maximum signal-to-noise ratio (S/N) at the target depth. Deeper targets usually require larger charge sizes. However, one should not choose a charge size that causes shotholes to blowout, because ejecting mud and wire out of the shothole loses a high percentage of the energy. A blowout can be dangerous because the drill hole acts like the barrel of a gun and concentrates the energy released. In addition, the detonator wire is usually ejected from the hole, and this wire can short out overhead power and telephone lines. Generally, larger charge sizes cause more ground roll and air blast contamination of the record (Figure 6.18). Alternatively, smaller charge sizes mean higher frequency content, but less energy going into the ground. Recently companies are using larger charge sizes again to put more energy into the ground. The overall frequency response can be higher than for smaller charge sizes, although the monitor records may appear to have a lower dominant frequency content (Figure 6.19). Deconvolution enhances

112 Source Equipment

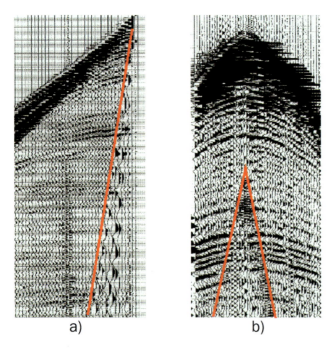

Fig. 6.18. Monitors showing (a) low-frequency ground roll and (b) high-frequency air blast.

the frequency content such that the bandwidth will be higher and have an improved S/N compared to a record with a smaller charge size.

The charge depth depends on the depth of the weathering layer and the level of noise interference one encounters when testing. Generally, the shallower the source (e.g., 6 m or *20 ft*), the stronger the air-blast and the ground roll. On the other hand, it is usually not economical to go much beyond 50 m *(160 ft)* depth. If the drilling is really tough and expensive (e.g., basalts or carbonates at surface), one may have

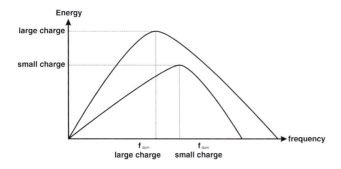

Fig. 6.19. Comparison of small and large charge.

to limit the shothole depth to as little as 2 m *(6 ft)*, or a surface shot may be used instead.

The number of holes per source location depends on the hole depth that one can drill economically and the apparent noise on single-hole records. Multiple-hole patterns may be required for sufficient energy penetration into the ground if difficult drilling dictates that shallow shotholes be used, or for noise cancellation purposes. In 3-D operations a source array will often be oriented in the same direction as the source lines for best noise cancellation, because receiver arrays are generally laid out in-line with the receiver line direction (See Chapter 8—Arrays).

The source type controls the sharpness and duration of the seismic impulse. Explosives are measured in terms of velocity of propagation, or the speed at which the shock wave propagates through the explovisve material. On one extreme, there is Pentolite, a very quick explosive with a detonation velocity of 8300 m/s *(27 000 ft/s)* that tends to produce a very sharp, short-duration impulse rich in high frequencies. In the middle of the velocity spectrum are seisgel and emulsion-based sources such as 60% Seis-gel. These sources have a velocity of about 5200 m/s *(17 000 ft/s)*, and tend to give a slightly broader wavelet than Pentolite. The main advantage of the gels over the Pentolite is the quick degradation of the charge to inert matter in case of a misfired shot. On the slow end of the spectrum is Primacord, which can detonate as slowly as 1500 m/s *(5000 ft/s)* and creates

Table 6.1. Typical dynamite test sequence.

Test No.	Charge Size		Depth		No. of Holes
	kg	lb	m	ft	
Test for charge size:					
1	0.5	*1*	15	*45*	1
2	1	*2*	15	*45*	1
3	2	*5*	15	*45*	1
4	4	*10*	15	*45*	1
Test for depth:					
5	1	*2*	6	*20*	1
6	1	*2*	9	*30*	1
Test for number of holes:					
7	1	*2*	15	*50*	3
8	1	*2*	15	*50*	5

a long duration explosion. These different explosives form a variety of possible sources that can be used in the field and only careful testing can tell which one is best for each survey.

6.2 DYNAMITE TESTING

Unless acquisition parameters are well known, a test sequence similar to the one in Table 6.1 is recommended. The test program should be located 30 to 40 stations from a BOL or EOL (beginning/end of line). Charge sizes are in kg *(lb)*. The test results should be delivered to the processor as soon as possible after recording so that the results can be reviewed in detail. The testing will aid in determining what source parameters should be used for the present program or any future surveys. These test parameters may vary greatly from area to area, and therefore, one should assure that the test sequence is suitable for the area under investigation.

6.3 DYNAMITE SHOOTING STRATEGY

A major advantage of dynamite operations is that several shooting crews can operate in tandem, shooting alternately (Figure 6.20). This technique allows one shooting crew to move to the next source location while the other shooting crew is making the electronic connection between the blaster and the dynamite wire and taking the shot. In dynamite operations it is especially important that vehicles "steady-up" while the recorder is in the listen mode following the shot. Cars or trucks traveling close to the line spread need to stop and shut off their engines to reduce the noise on the recorded data.

6.4 VIBRATORS

Vertical vibrators (Figure 6.21) produce an asymmetric radiation pattern of *P*-waves and *S*-waves. Horizontal vibrators produce weak *P*-waves and ro-

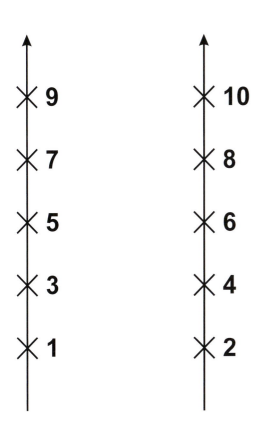

Fig. 6.20. Tandem shooting on two parallel source lines.

Fig. 6.21. Vibrator base plate.

Fig. 6.22. Vibrator—buggy mounted.

Fig. 6.24. Vibrator—truck mounted.

Fig. 6.23. Vibrator transport.

bust *S*-waves. If multiple dynamite patterns do not put enough energy into the ground, vibrators may be preferred on technical grounds, regardless of relative cost. For example, one may encounter poor dynamite data in an area with 200 m *(650 ft)* of glacial till, but still obtain outstanding data when a Vibroseis source is used.

Vibrators are designed in two basic groups: buggy-mounted and truck-mounted units. The buggy vibrator (Figure 6.22) is usually more mobile and causes less damage off-road than a truck-mounted vibrator. Most buggy vibrators are pivoted, or swivel behind the cab, and many can also pivot vertically behind the cab. This maneuverability and flexibility allows for less environmental impact and improved ease of use. On the downside, buggy vibrators must be trucked (Figure 6.23) and can get stuck in fine sand. The truck-mounted vibrator (Figure 6.24) is more efficient for roadwork and use around more developed areas. The truck-mounted vibrator does not perform as well off-road as a buggy, so vibrator choice is usually dictated by the project logistics.

6.5 VIBRATOR ARRAY CONCEPTS

In vibrator acquisition, the source effort depends on the following parameters:

sweep length
number of sweeps
number of vibrators
fundamental ground force

The sweep length (time) is usually in the range of 4 to 20 s. Testing will determine the combination of sweep length with the other acquisition parameters that offer the best results. The longer the sweep length, the more time the vibrators spend putting certain frequency ranges into the ground. A long sweep can be good from an energy/bandwidth point of view and also bad because long sweeps can sometimes create significant ground roll and other noise.

The number of sweeps summed at each source station may range between 4 and 20. Noise cancellation is improved by increasing the number of sweeps. For example, wind noise on far traces can be reduced significantly with a higher number of sweeps.

The product of the sweep length and the number of sweeps is referred to as the pad time and is one of the most important factors in determining the crew cost:

$$\text{pad time} = (\text{sweep length}) \times (\text{number of sweeps})$$
(6.1)

Pad time usually varies between 64 and 144 s per source point; however, under certain circumstances, it may be as low as 10 s. Generally, it is more economical to decrease the number of sweeps because that reduces the vibrator moves. Fewer and longer sweeps are more desirable than are increased numbers of short sweeps.

The number of vibrators usually varies between 3 and 5 per array (Figure 6.25), and most crews are running two vibrator arrays per acquisition crew. Acquiring data with single or dual vibrators will not offer the noise cancellation that can be accomplished with a larger number of vibrators. Because vibrator data almost always use source arrays, one should always consider using receiver arrays (see Chapter 8) on vibrator surveys.

Equation 6.2 gives the peak signal to RMS noise of the correlated wavelet improvement in dB. Variations in the number of vibrators and changes in the fundamental ground force affect the S/N improvement more than the sweep length, and the number of sweeps assuming that the bandwidth of the noise (BW; in Hz) does not change very much (Malcolm Lansley, pers. comm.).

S/N improvement in dB = $20 \log_{10}$ (number of vibrators × fundamental ground force × (sweep length × number of sweeps × BW)$^{1/2}$) (6.2)

Vibrators can be spread over a great distance to create a source pattern, and many sweeps can be summed to create a large source. A continuous succession of pad locations results (Figure 6.26). The distance from the pad of the first vibrator to the pad of the last vibrator determines the array length. This array may be stationary or can be moved up over a predetermined distance about the source point as shown in Figure 6.27. The total source array length is

effective source array length = array length + total move-up = number of vibrators × pad distance + number of move-ups × move-up distance. (6.3)

Fig. 6.25. Vibrator array.

Example: effective source array length =
4 × 10 m + 3 × 10 m = 70 m

or

effective source array length =
4 × 41 ft + 3 × 41 ft = 288 ft.

Fig. 6.26. Vibrator pad marks.

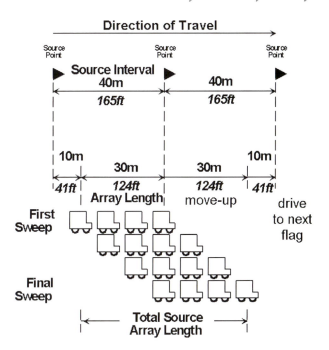

Fig. 6.27. Vibrator array, move-up, and drag.

Move-up attenuates noise and is easier on operators, equipment, and the environment. A small amount of move-up may decrease air blast noise and ground roll at near offsets. Larger move-ups may increase multiples and should be avoided.

Often the first vibrator sits near the source flag for the first sweep. The vibrator array then moves up between each sweep so that the last vibrator sits near the flag for the final sweep. This vibrator movement creates a weighted array (in Figure 6.27 the weights are 1, 2, 3, 4, 3, 2, 1). If, after the final sweep, the first vibrator is not at the next flag, the entire array drives up to the next flag.

The changes in vibrator coupling that occur while using a move-up can be neglected, because the vibrator coupling changes from sweep to sweep and even within a sweep from beginning to end.

6.6 VIBRATOR TESTING

Similarity tests ("sims") are performed at least twice daily to confirm that vibrators are performing satisfactorily. Hard-wire tests, with a physical cable connection from the recorder to the vibrators, are the most thorough performance check and these should

Table 6.2. Typical vibrator test sequence.

Test no.	No. of vibrators	Sweep frequency (Hz)	Number of sweeps	Sweep length (s)	Dwell dB/octave	Array length (m)	Move-up (m)	Total array (m)
Test for sweep frequency								
1	4	10–80	8	12	3	30	30	60
2	**4**	**10–90**	**8**	**12**	**3**	**30**	**30**	**60**
3	4	10–100	8	12	3	30	30	60
4	4	10–110	8	12	3	30	30	60
5	4	10–120	8	12	3	30	30	60
Test for dwell								
6	4	10–90	8	12	none	30	30	60
7	4	10–90	8	12	6	30	30	60
Test for drag length								
8	4	10–90	8	12	3	30	10	40
9	4	10–90	8	12	3	30	20	50
10	4	10–90	8	12	3	30	30	60
Test for number of sweeps								
11	4	10–90	1	12	3	30	30	60
12	4	10–90	2	12	3	30	30	60
13	4	10–90	4	12	3	30	30	60
14	4	10–90	16	12	3	30	30	60
15	4	10–90	32	12	3	30	30	60
Test for sweep length								
16	4	10–90	8	4	3	30	30	60
17	4	10–90	8	8	3	30	30	60
18	4	10–90	8	16	3	30	30	60
19	4	10–90	8	32	3	30	30	60
Test for number of vibrators								
20	1	10–90	8	12	3	0	0	0
21	2	10–90	8	12	3	10	0	10
22	3	10–90	8	12	3	20	0	20
23	4	10–90	8	12	3	30	0	30

be completed at the start of the job and anytime equipment is pulled offline for maintenance or repair. Vibrator testing usually requires a significant amount of time in the field to record different kinds of source parameters; e.g., varying the sweep range, dwell, drag length, number of sweeps, sweep length, boost or gain percentages, peak force, and number of vibrators. Unless acquisition parameters are well known, a test sequence similar to the one in Table 6.2 is recommended. A standard sweep might be that indicated as test no. 2. The type of test sequence may vary greatly from area to area. The S/N formula can be used to design a test sequence so that the parameters vary sufficiently to show S/N variations that are easily visible. In Table 6.2 this requirement has been incorporated by increasing the number of sweeps and the sweep length by factors of 2. The improvement in the S/N should be 3 dB with each doubling of effort [see equation (6.2)].

It is important for vibrator testing to have a field processing system that will allow the geophysicist to make on-the-spot decisions about further testing that may be necessary beyond an initial test sequence. Band-pass filters of deconvolved shot records give a good indication of the necessary sweep range. The quality control geophysicist must ensure that the field processing system can interface with the recording system. The most basic questions in this age of changing recording media are: Are the tape drives the same; how long does it take to transfer the information; and is that time reasonable? It is recommended that once basic parameters are picked, the first few square kilometers of the survey are fast-tracked, processed in the field, and the results analyzed by the interpreters for data usability and quality (Duncan et al., 1996). One should retain the test results to reduce the amount of testing necessary during future programs in the area.

Vibrators can perform sweeps in a wide variety of options. Down-sweeps that were in vogue in the 1970s have been replaced by up-sweeps. Up-sweeps start at a low frequency of about 10 Hz and sweep to 80, 100, 120 Hz, or more. Sweeps may be linear, which means an equal amount of time is spent sweeping through the frequency ranges, or a dwell or "boost" can be incorporated into the system to allow enhancement of higher frequencies (Figure 6.28). Such a dwell creates nonlinear sweeps. Some dwell is useful in increasing the high-frequency range of the final stacked seismic section. However, too much dwell can create noisy record sections due to lack of low-frequency data needed for interpretability. The normal dwell range is between 3 and 6 dB/octave. Careful testing

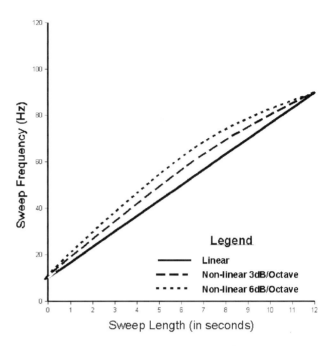

Fig. 6.28. Linear versus nonlinear sweeps.

and design of dwell ensures an optimum S/N (Pritchett, 1994). The choice between linear and nonlinear sweeps usually does not affect cost as long as the pad time remains the same. Vari-sweeps can be used to enhance certain frequencies beyond the dwell in a selective manner. In this technique, several sweep ranges are selected to span narrow frequency bands and then summed.

For vibrators, a sweep rate is established as follows, and this parameter has to be set in the recording truck as well as in the vibrators:

$$\text{sweep rate} = \frac{\text{highest frequency} - \text{lowest frequency}}{\text{sweep length}}.$$

(6.4)

One needs to perform phase, force, and frequency versus time plots for all vibrators before and after testing to evaluate their performance. Figure 6.29 shows the performance of four vibrators. Three lines are solid to signify the similarity of those performances. The poorly performing vibrator is noted as a dashed line. The phase versus time plot should indicate essentially a constant phase after 1 to 2 s. Any major deviation (i.e., $>5°$) is unacceptable, as is the one vibrator that shows a somewhat linear change in phase over time (dashed line).

118 *Source Equipment*

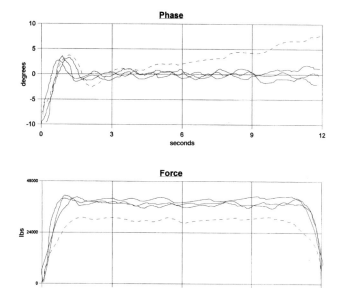

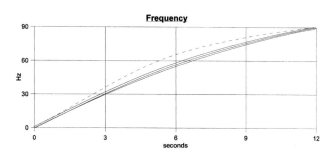

Fig. 6.29. Phase, force, and frequency versus time quality-control plots.

Vibrators should operate at 70 to 90% of peak force. Each vibrator performs more consistently by operating at such a percentage rather than closer to peak force. Also, operating at a higher force introduces larger vibrator-to-vibrator differences, which results in poorer quality data. On the other hand, the vibrator force should not drop below 85% of the operating force throughout the sweep (e.g., dashed line). The frequency versus time plot in Figure 6.29 indicates a nonlinear sweep with three vibrators having similar sweeps, while one vibrator is sweeping with a higher dwell.

The vibrator-sourced data can be recorded in two ways. In normal practice, records are correlated with the vibrator source signal before putting the records to tape. Recording uncorrelated records increases tape requirements by a large amount—a factor of five in this example (sweep length of 12 s + the listen time of 3 s, i.e., 15 s).

6.7 VIBRATOR DEPLOYMENT STRATEGY

For large surveys, one may want to employ two sets of four or five vibrators to allow more effective data acquisition. This source strategy increases the daily cost but reduces the amount of recording time. The use of several sets of vibrators depends on several factors, such as the terrain, required sweep effort, and drive time between source points. The fifth vibrator in each set can be an in-the-field spare so production is not lost if a vibrator needs repair. Such contingency planning is essential, particularly in remote areas.

Vibrator surveys have the major advantages that source points may be occupied repeatedly to increase the fold after having moved the patch, or to repeat a source point. This flexibility is much more difficult to achieve with dynamite programs because of shothole drilling time, patch moves, and confusion over which holes are to be shot. During vibrator operations, line crews can still lay out and pick up cables and phones. This crew movement usually cannot take place while recording dynamite data.

When source points need to be occupied repeatedly, it is important to strive for nonduplicating source-to-receiver raypaths. This also applies to the recovery of make-up source points. Recovery-source-point positioning is also better when the source points are perpendicular to the source lines rather than parallel to the line. This station movement reduces disturbances in the fold distribution.

Slip-sweep acquisition has been instrumental in reducing the time spent on vibrator source points. This is performed by squeezing more VPs into a given amount of recording time by letting subsequent sweeps overlap by a certain extent (Rozemond, 1996).

Some Vibroseis surveys are performed with single sweeps at source stations a short distance apart (e.g., every 10 m). The data from each of these "sweeps" is correlated separately. During processing, these "shots" are treated individually, which leads to folds as high as 720. The data resulting from such surveys is generally of extremely high quality. To date, most of these types of surveys have been conducted in the Middle East.

Allen et al. (1998) presented a new method of high-fidelity recording using vibrators which includes the simultaneous recording of data from several vibrators. Processing such data uses measured motion signals on vibrators to separate the data for each vibrator and has resulted in higher fold data with a significant improvement in the data quality (Wilkinson et al., 1998).

6.8 OTHER SOURCES

Although dynamite and vibroseis are used on the great majority of surveys, other sources can be and are used in the field for 3-D surveys, such as:

air guns and mud guns (used in transition zone surveys)
shotgun (Betsy)
Mini-Sosie (thumper)
land air-gun
Dinoseis
Elastic Wave Generator—EWG (Sercel)
Mini-Vibs

It is possible to use one of these (environmentally noninvasive) surface sources in certain parts of the 3-D survey to collect near-offset (shallow) data where it is impossible to use dynamite or Vibroseis. Heavier (stronger) sources can collect long-offset (deep) data by undershooting those parts. The processing center will have the unenviable task of matching the phase of the shallow and deep data in order to create one data set.

Chapter 6 Quiz

1. What parameters determine the effort in dynamite shooting?
2. What is meant by the expression "pad time"?
3. Explain the terms array length and move-up by drawing a diagram.

7

Recording Equipment

7.1 RECEIVERS

The receiver type depends on the characteristics of the data to be recorded and the environment where the data acquisition occurs. A variety of receivers and their uses are listed in Table 7.1. In normal land operations, geophones have a resonant frequency of 10 or 14 Hz, but in some parts of the world it is still normal practice to use 6- or 8-Hz phones. However, geophones with resonant frequencies up to 40 Hz are being manufactured (Figure 7.1). Receivers are usually wired in groups of 1, 4, 6, 9, 12, or 24 (Figure 7.2). While the trend is towards higher numbers of phones (9, 12, 24—or even 72 in the Middle East), lower numbers (e.g., 6) are still used in certain areas, e.g., South America. If the takeouts on the receiver line cables are spaced farther apart than the group interval, the cables can be coiled. Geophone stations may be added as single stations (telemetry systems) or in sets of 4 or 6 groups (distributed systems), depending on the recording instrumentation employed. Generally each geophone group can be addressed individually from the recorder.

New 3-component solid-state receivers are under development. These sensors will have a digital output with a flat spectrum. The micromachined accelerometers utilize sigma-delta technology and provide high resolution and ultralow distortion. Since three sensors are aligned orthogonal to each other, the deployment angle can be measured directly and any tilt of the sensor unit is immaterial.

Future sensor developments should include a significant weight reduction, allowing much higher channel capacities to be deployed in less time. In-line sampling might be much closer in the future, with smaller bins being used and higher fold data being recorded.

Receiver arrays are formed by spreading receivers over some distance (e.g., 12 geophones over 20 m *or 40 to 55 ft*). For operational reasons, the groups are laid out in-line with the recording spread, cross-line, or in a pattern such as a circle or a box, around the station flag (e.g., Figure 7.3). Little work is available in the public domain about the use of receiver arrays in 3-D seismic recording (Regone, 1994). A box array improves the two-dimensional effect of the array, and the linear dimension of the array is reduced. Clustering of some geophones may create a weighted spread, but reduces the effectiveness of long-wavelength noise attenuation. A more complete discussion of arrays is provided in Chapter 8.

In hilly terrain, where the height difference between the ends of any receiver group exceeds 2 m *or 6 ft*, geophones may be clustered in a small area. In steep terrain (over 5 m *or 15 ft* elevation difference) one can spread the phones out parallel to topographic contours to minimize interarray statics smear. The geophones may need to be temporarily buried to reduce wind noise. For a 3-D survey that may need to be reshot at a later date (e.g., fire-flood monitoring), it may be best to bury or cement phones permanently in the ground.

Three-component 3-D recording requires three times the number of channels of recording capacity since each component is recorded separately (Figure 7.4, center geophone). This increased number of channels may make it difficult to create a patch that creates

Table 7.1. Types of receivers.

Type	Resonant Frequency	Environment
geophone	6, 8, 10–40 Hz	land
marsh phones	10–14 Hz	transition zone
hydrophones	10–14 Hz	marine
3-component	2–8 Hz	shear wave

122 *Recording Equipment*

Fig. 7.1. Geophone.

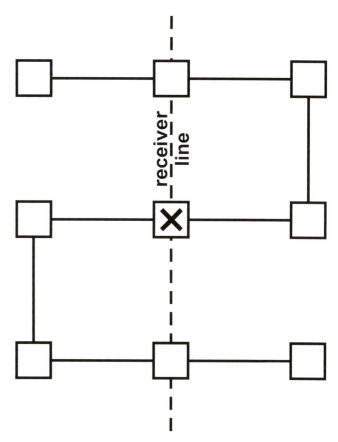

Fig. 7.3. Nine-element box array. Each square is a geophone. X is the station flag.

Fig. 7.2. String of geophones on a hoop and cables used in seismic acquisition.

Fig. 7.4. Different geophone types.

sufficient fold. Since shear-wave reflections contain a lower frequency bandwidth, phones with lower resonant frequencies are used.

In transition zone environments (shallow marine, marsh, or lake), one may want to use hydrophones or marsh phones. Marsh phones should be pole planted to secure the best coupling. Hydrophones can be strung out in a pattern on the lake bottom. Often the available equipment will allow only one receiver element per group. One has to carefully examine the effect that a one-element array will have on noise reduction. Hydrophones are pressure sensitive rather than velocity sensitive as geophones are, and reflected pressure wavefields are 180° out of phase with reflected particle-velocity wavefields. The timing relationship between the primary pressure reflection and its ghost causes an apparent 90° phase shift between hydrophones relative to geophones at shallow depths near the shoreline. Crosscorrelation filters can be designed to obtain the best phase match between different recording instruments and source wavelets.

Once sufficient receiver groups are deployed to record the first source point, the operator in the recording truck needs to stabilize the spreads by checking for proper cable connections electronically. Should there be any poor connections or leakages, the line crews need to check, repair, and (if necessary) replace spread cables, boxes, batteries, and connections. The operator should avoid a total system shutdown once the spread is stable, because a wake-up across the spread involves rechecking the entire spread again, which is rather time consuming (another reason for 24-hour operation). However, radio telemetry systems do not suffer from such problems and fast wake up times are normal because all boxes wake up simultaneously rather than with a cable system where a bad box can cause the remainder of the spread past the bad box to be inactive.

7.2 RECORDERS

There is a large variety of recording equipment available for 3-D surveys. Today, many acquisition systems provide 24-bit recording technology; most of the systems listed in Table 7.2 do so. A 24-bit technology system offers high fidelity because it records data over a large dynamic range. The listing in Table 7.2

Table 7.2. 3-D Recording instrumentation.*

Manufacturer	System	T/D/R	Boxes	Stations per box	Line units	Central system
Fairfield	Telseis Star	T	RU	1–6	directly to CRS/CRS	CRS
Fairfield	BOX	T/D/R	RU	1–8		CRS
Geo-X	Aram	D	RAM	8	LTU	CRU
Input/Output	I/O I	D	RSC	6	LTU	SCM
Input/Output	I/O II	D	MRX	6	ALX	SCM
Input/Output	RSR	R	RSR	6		
Input/Output	System 2000	D	MRX	6	ALX	SCM
JAPEX	G.DAPS-4					
Sercel	Eagle	T	SAR	1–6	directly to CRS	CRS
Sercel	SN388	D	SU	1–6	CSU	CCU
Sercel	408 UL	T/D/R	FDU	1	LAUX	CM
Syntron	Polyseis	T	RTU			

D = Distributed system   T = Telemetry system   R = Remote seismic recording

ALX	Advance line tap	LTU	Line tap unit
CCU	Central control unit	MRX	Miniature remote signal conditioner, extended range
CLU	Cross-line unit		
CMM	Central management module	RAM	Remote acquisition module
CRS	Central recording system	RSC	Remote signal conditioner
CRU	Central recording unit	RTU	Radio telemetry unit
CSU	Crossing station unit	RU	Remote unit
FDU	Field data unit	SAR	Seismic acquisition remote unit
LAM	Line acquisition module	SCM	System control module
LIM	Line interface module	SU	Station unit

*For further specifications please refer directly to the relevant company's literature.

is not an endorsement of particular instruments but merely a summary of some technical specifications. Peculiarities for each system need to be examined for the task at hand, e.g., I/O systems need an additional LIM when recording more than 1015 channels (1024 minus the auxiliary channels). In land operations, these recording units are usually truck or buggy mounted and can, therefore, travel easily to areas of data acquisition.

Lower channel count systems with higher sampling rates, such as the DMT/SUMMIT and the 24-bit OYO DAS, can be used for small, near-surface 3-D surveys. In the case of very low channel count systems (e.g., less than 120), it is normal for several recorders to be used together in a master-slave system to reach sufficient channel capacity even for small 3-D surveys. There are still a number of 16-bit technology recorders, but the preference is always to use a 24-bit system. Finally, older analog instruments like the DFS V are no longer used for acquiring 3-D surveys.

Temporal sampling in 3-D geometry follows the sample theory of 2-D. Usually, one records seismic data at a 2 ms sample rate with a high-cut of 128 Hz as an anti-alias filter. Resampling to 4 ms is an option to reduce the data volume before processing, although there is a trend towards higher sample rates (e.g., 0.5 or 1 ms). Deeper targets rarely send back high frequencies, thus it is illogical to sample deep data (e.g., more than 2 s two-way time) at a rate less than 2 ms.

If a 3-D survey crosses a variety of terrains (e.g., mountain, plain, transition zone), it is desirable to use one type of recorder to cover all of the survey area. Thus shots of different types (dynamite, airgun, etc.) in the mountains or in the swamp can be recorded by the same instrument. If more than one recorder is used, amplitude and phase matching will be required to compensate for the recorder differences. "Seamless" receiver coverage from a variety of sources enables application of surface-consistent processes as deconvolution, statics, and amplitude correction. Present industry trends favor such "universal" recording systems.

7.3 DISTRIBUTED SYSTEMS

Distributed systems have cable connections between the units that record the field data and the control system in the recording truck. Usually, these connections are made via a line unit (Figure 7.5) that gathers the information from several different geo-

Fig. 7.5. Solar battery powered boxes.

phone groups. The recording equipment is set up in an arrangement similar to the one in Figure 7.6. In most distributed systems, some number of groups of geophones are connected to a box. These boxes convert the analog signal to digital form (samples). The boxes are normally wired together along a "branch line" that usually takes the form of a twisted pair. Branch lines feed into cross-line units, which in turn are connected by a "trunk line" and fed directly into the central recorder (Figure 7.7). In the case of Vibroseis, the information must be correlated and recorded to tape. This correlation may occur in the individual boxes or at the recording truck (Figure 7.8). These trucks tend to be very cramped with a correlator unit (Figure 7.9), camera units (Figure 7.10), tape drives, and computers that control operations.

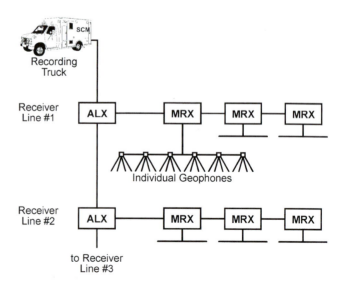

Fig. 7.6. Typical field setup for a distribution system (Input/Output System Two). ALX = advance line tap, MRX = miniature remote signal conditioner.

Fig. 7.7. System control unit.

Fig. 7.8. Recorder.

Fig. 7.9. Correlator unit.

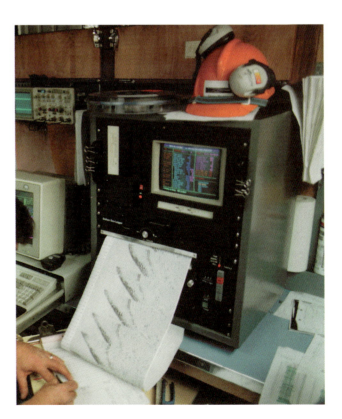

Fig. 7.10. Camera unit and monitor record.

7.4 TELEMETRY SYSTEMS

True telemetry systems have no physical connection between the station recording unit and the control system in the recording truck. These systems should be used where access is limited due to rugged terrain, permit problems, or any other reason. A typical field setup is shown in Figure 7.11. Again for convenience and cross-reference to Table 7.2, the Sercel Eagle system is illustrated in this example. The SAR (Seismic acquisition remote unit) records the signal and sends it via radio frequencies to the CRS (Central recording station). Telemetry systems have considerable advantages over distributed systems, such as:

Reduced HSE (Health, Safety, Environment) impact:

> No damage to cable from wildlife — or the environment (e.g., oyster beds)
> No cable to carry in rugged (and therefore dangerous) terrain
> Arbitrary spacing is not a problem

Quicker setup time:

> Each box is independent of other boxes; therefore, shooting can normally be started much quicker than with a cable system where one bad box can make the rest of the spread inactive.

Better production:

> Acquisition does not have to stop because of one or two bad boxes.

Some telemetry systems (Fairfield BOX and Syntron Polyseis) can receive data in real time. Other telemetry systems have a disadvantage over distributed systems in that the radio transmission of the data from the

Fig. 7.12. Antenna set-up for telemetry operations.

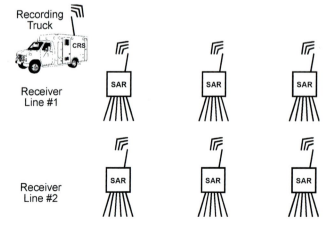

Fig. 7.11. Typical field setup for a telemetry system (Sercel Eagle). SAR = seismic acquisition remote unit.

Fig. 7.13. Remote boxes.

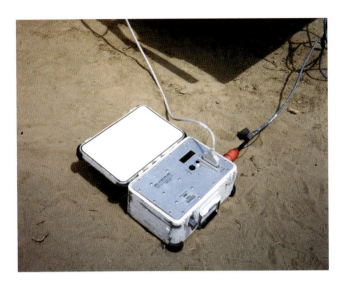

Fig. 7.14. Data collection unit.

Fig. 7.15. Data transcriber.

boxes to the recording unit takes longer than real time. For some systems, data transmission time may be on the order of minutes per source point, which may slow down the shooting crew. Tree cover may also cause a problem for the signal transmission, and FM interference may be significant in populated areas. Mixed systems (some telemetry, some distributed) may be used to cross rivers or roads at select locations.

7.5 REMOTE STORAGE

Some systems have been developed that allow the storage of many records in the box near the geophone groups. Examples of these systems are the Fairfield BOX and I/O RSR system. The RSR system is composed of a recording truck, a large antenna (Figure 7.12), and individual remote seismic recording units or "boxes" (Figure 7.13). The data are recorded in the field and stored in the BOX or RSR box (recordings from several hundred or more source points) and then collected a few times per week depending on how much memory is available. The data collection unit or DCU (Figure 7.14) is temporarily connected to the (BOX or RSR) box and then the stored seismic information is downloaded to the DCU. After data transfer, the memory is erased in the (BOX or RSR) box for further use. The DCU is then disconnected from the BOX or RSR and connected to the transcriber and correlator unit (Figure 7.15) for further transcription. The transcriber then sorts the data to shot records, correlates out the pilot sweep, and writes the data to a tape for permanent storage. In some cases (e.g., Fairfield BOX), the stored data can be collected by radio, once the central recorder comes within radio range of the remote storage units. Such a data collection method can enormously reduce the HSE impact.

Data QC is difficult or impossible in some cases with remote storage. Ideally, seismic trace attributes should be available for all the traces of a shot shortly after the shot is taken. These attributes (e.g., S/N, rms energy) provide reassurance that good seismic data are being recorded. Some remote storage systems can provide this trace attribute information via radio for each shot.

Chapter 7 Quiz

1. What type of receivers should be considered for use in transition zone environments?
2. What is involved in stabilizing the spread?
3. Describe the difference between distributed and telemetry systems.
4. How does a remote seismic recording system work?

8

Arrays

8.1 THE QUESTION OF ARRAYS

Array design in 3-D surveys requires careful consideration. A good treatment of arrays in 2-D surveys can be found in Vermeer (1990). If a receiver array is used, it should be as simple as possible. Circles, box patterns, or in-line arrays are commonly used today because of the ease and efficiency of their layout. A designer may place a major emphasis on an intricate receiver array, but this message may be hard to convey to the field personnel who may not understand the necessity of carrying out instructions accurately. In such cases, the designer must work with the field personnel directly to achieve meticulous placement of the arrays.

In 3-D surveys, source energy arrives from many directions. The array response varies dramatically depending on the azimuth between point sources (e.g., dynamite) and receiver arrays. If geophone arrays are laid in-line, the response from a point source fired from an in-line position will be attenuated, while a broadside source into such an array will not be affected by the array. Because of this variation, many companies prefer no arrays at all when recording 3-D data with single-hole dynamite source points, and use omnidirectional arrays such as circles or clustered arrays.

Field tests for linear versus podded geophone arrays have indicated that there may be situations where arrays can attenuate reflections more than groundroll does (Wittick, 1998). The tests included in-line as well as cross-line recording, i.e., the wavefields hit the geophone groups linearly as well as broadside.

Regone (1998) has shown that dense wavefield sampling through the use of areal grids of receivers is effective in reducing scattered noise. Direct-arrival noise requires the use of adequate spatial sampling (and usually field arrays), while ambient noise is best overcome by an appropriate choice of source type, source strength, and fold.

Arrays in the source line direction should complement arrays in the receiver line direction as much as possible (Vermeer, 1998a). Figure 8.1 shows the different array combinations that may be considered. Figure 8.1a is a single source point being recorded by a linear receiver array, the series of midpoints indicates possible smearing in the receiver line direction. Figure 8.1b shows the areal midpoint sampling that might happen with a combination of linear source and receiver arrays. This areal midpoint coverage is summed as one trace. Figure 8.1c indicates the midpoint coverage that can be achieved with a single source and an areal receiver array. The greater the number of elements in a receiver array, the larger is the areal midpoint coverage. Such areal arrays are effective in reducing back-scattered noise.

8.2 GEOPHONE ARRAYS

Figure 8.2 illustrates a five-element geophone array response oriented east-west to energy arriving from various directions. Figure 8.2a shows the array response in a linear plot, while Figure 8.2b shows the response in a radial presentation. Note the poor attenuation of broadside energy, at 90° (north) and 270° (south), and good attenuation at 0° (east) and 180° (west). Figure 8.2b shows the 2-D array response with 0° being in-line to the geophones, i.e., east-west. A cross-sectional display of the array response is shown at the bottom of each figure, e.g. at 209.8° in Figure 8.2a. The location of each profile view is indicated by a thick black line through each 2-D response.

130   Arrays

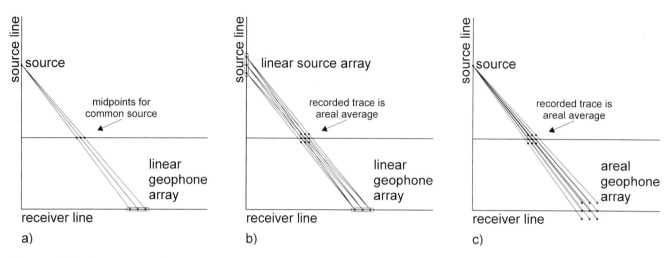

Fig. 8.1. Midpoint coverage for: a. a single source into a linear geophone array (after Vermeer, 1998a), b. a linear source array into a linear geophone array, and c. a single source into an areal geophone array.

8.3 SOURCE ARRAYS

Figure 8.3 shows an array response of a simulated four-vibrator drag in the north-south direction. Note the good attenuation at 90° (north) and 270° (south) and poor attenuation, or full response, at 0° (east) and 180° (west). Typical patterns are four vibrators with 10 m between pads (40 m effective array length) moving up 5 m (or less) at a time. In the example of Figure 8.3a, the distance between pads is 5 m and move-up is 5 m, for a total of four sweeps. The effective array length is 35 m; the response of four vibrators is shown as a radial pattern in Figure 8.3b.

8.4 COMBINED ARRAY RESPONSE

Figure 8.4 indicates the combined response of the receiver and sources arrays. It is important to note that these graphs are the theoretical response in an ideal situation with the geophone arrays and source arrays arranged orthogonally. This particular array provides good noise attenuation in all directions. Each layout method has its own particular recording geometry. Accordingly, arrays will be laid out differently.

The theoretical response should be evaluated before proceeding with a particular design. The noise region to be attenuated is indicated from a wavelength of 0.1 to 0.3 in all array response displays.

8.5 STACK ARRAYS

Stack arrays have been discussed by Anstey (1986a, 1986b, 1986c, 1987, 1989). The 2-D stack array approach creates an even, continuous, uniform succession of geophones across the CMP gather. In land operations, split spreads are used with the following geometrical parameters imposed:

  group length is equal to the group interval,
  the source interval is equal to the group interval, and
  the source points are between the groups.

These stack arrays depend on the trace offsets from their respective source points. If isotropic noise is assumed, the 3-D stack array becomes a 2-D array formed by different offset traces. Thus, the response at each azimuth is the same. Because 3-D CMP bins always contain traces of differing offsets, the stack array

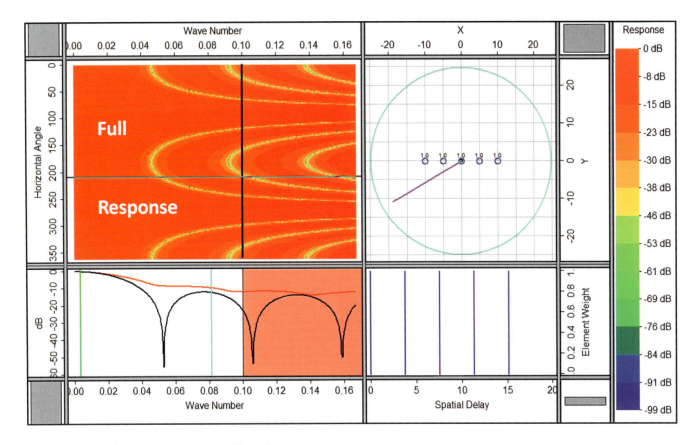

Fig. 8.2a. Geophone array response (linear).

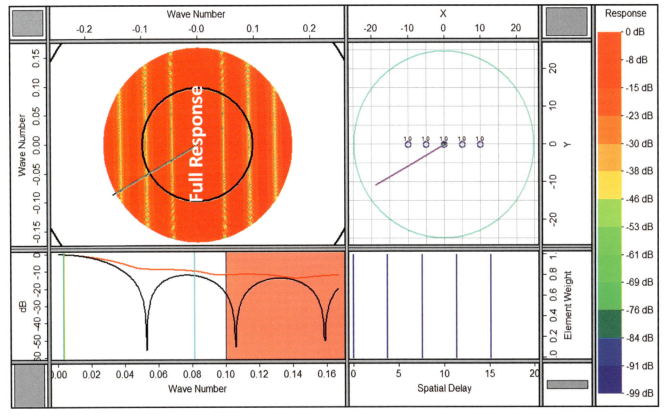

Fig. 8.2b. Geophone array response (radial).

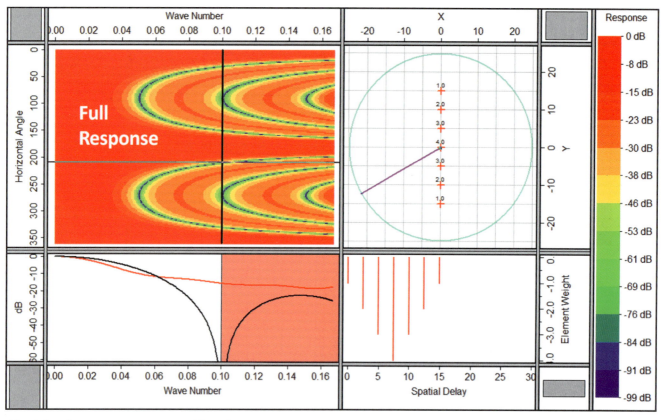

Fig. 8.3a. Source array response (linear).

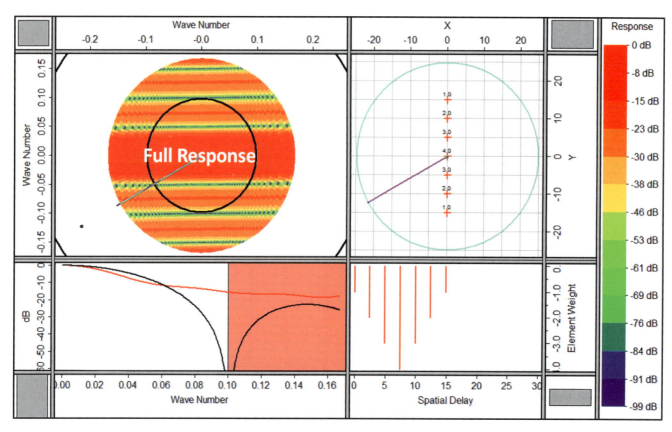

Fig. 8.3b. Source array response (radial).

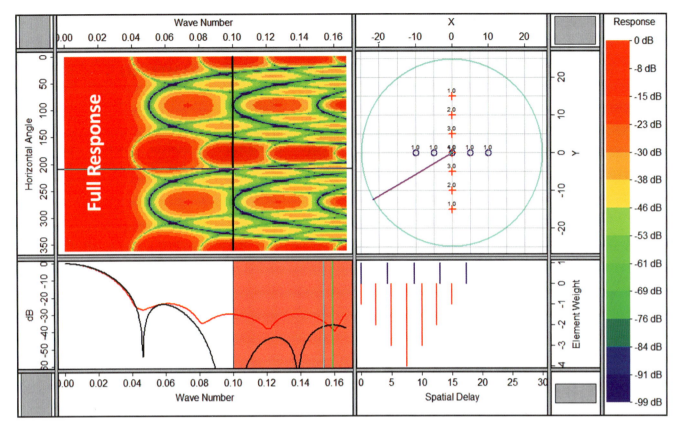

Fig. 8.4a. Combined 2-dimensional array response (linear).

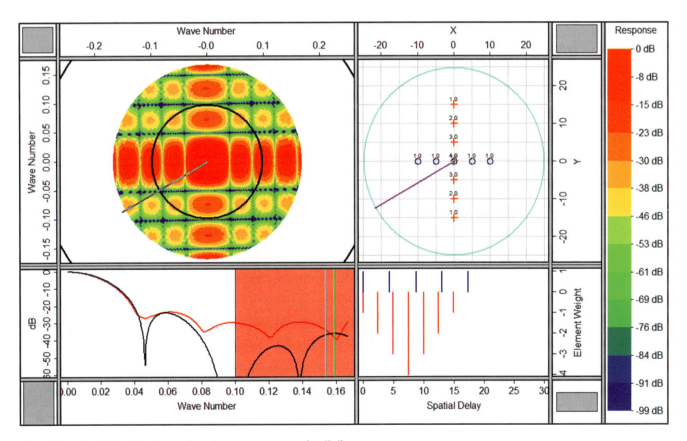

Fig. 8.4b. Combined 2-dimensional array response (radial).

effect is different in each CMP bin and is often negligible because of the irregular offset mix at each CMP.

8.6 HANDS-OFF ACQUISITION TECHNIQUE

Ongkiehong and Askin (1988) have taken the stack array approach one step further by imposing equal array lengths for the source and receiver patterns. For symmetric sampling, the number of elements in each array, whether source or receiver, should be the same. Their conclusion is that *any deviation from homogeneity in the fundamental sampling operator is ultimately detrimental.* In 3-D field acquisition, such symmetric sampling may not be economically possible, and some compromises need to be made.

8.7 SYMMETRIC SAMPLING

Vermeer (1998a) supports the idea of symmetric sampling as well. Anything that is done on the receiver side should be done on the source side as well. Hence the source and receiver station as well as line intervals are equal. The maximum offset is measured in the in-line direction and it is equal to the maximum offset measured in the cross-line direction. This results in equal in-line and cross-line fold also. Any receiver arrays are mirrored as source arrays. As mentioned in the previous section this is an ideal scenario, which requires compromises in practical situations.

9

Practical Field Considerations

9.1 SURVEYING

The designer of the 3-D seismic program and the surveyors must maintain close communication in order to locate the program correctly and execute it properly. The designer needs to pass on exact instructions for allowances that can be made for any anticipated changes, e.g., how make-up source points are to be located. In steep terrain, surveyors often assume that distance can be measured along the chain (slope chaining), instead of horizontally. Chaining horizontally is the only effective way of ensuring that a central midpoint distribution is maintained (Figure 9.1).

Modern Global Positioning System (GPS) surveying has significantly changed the speed of layout and the precision of acquired survey data in rough terrain. GPS units are usually carried in a backpack (Figure 9.2). A hand-held data collector is used to navigate to the next survey point or the unit can be mounted on an all terrain vehicle (quad) (Figure 9.3) for faster field operations. Many acquisition contractors put regulators on the quads to limit the speed of the vehicle for safety reasons. Along the route to the survey point, the line is flagged and marked for the layout crew to follow (Figure 9.4).

Vegetation (especially large wet trees) can severely impede GPS signal reception. Recently inertial navigation systems that can be carried in a backpack have become available. These have been used in dense jungles with considerable success.

A 3-D seismic design generally goes through three mapping stages: the theoretical plot, the preplot, and the final plan. The theoretical plot is merely an indication of the theoretical station and line locations (Figure 9.5a). A preplot of the program at a working scale gives the surveyors a good basis to work from and gives the designer the comfort that the anticipated location for every source and receiver point has been documented (Figure 9.5b). One should number the source and receiver locations in such a manner that no two locations have the same number. Topography, wells, buildings, pipelines, existing seismic lines, and other surface culture have an effect on the location of source and receiver lines and stations (Table 9.1). The designer takes these into account as much as possible in the planning stage, especially topographical restrictions. It is helpful to have the preplot registered with local coordinates and possibly overlain on an areal photograph. The surveyor has to provide detailed information regarding exclusions, skids, and offsets to the designer, who then decides on the appropriateness of the changes and may possibly remodel the program. The designer must indicate how far sources or receivers may be moved before they should be dropped entirely. Generally, offsetting stations by more than the line interval is unacceptable. Where complex surface obstacles exist, it might be beneficial to have a designer in the field with the surveyors. Many problems can be solved most easily on the spot. Ideally this advance person has seismic field experience and is equipped with a laptop computer with the initial 3-D design loaded. The surveyor must establish the format of the digital survey information prior to the field visit. The SEG-P1 format is widely accepted for this purpose. Standard exchange formats for positional information have been published by the Society of Exploration Geophysicists (1983). Electronic data transfer of survey information can significantly reduce the likelihood of human error when copying data.

The final plan uses the actual survey information (Figure 9.5c). The final survey plan of the 3-D program may not bear much resemblance to the preplot, but it is still an essential element of the surveying

Fig. 9.1. Chaining.

Fig. 9.2. Backpack GPS unit.

Fig. 9.3. Quad with speed regulator.

Fig. 9.4. Survey flags along seismic line.

9.1 Surveying   137

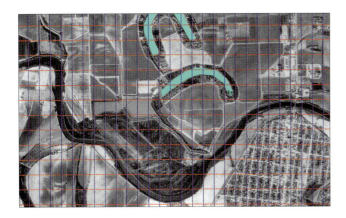

Fig. 9.5a. Theoretical layout.

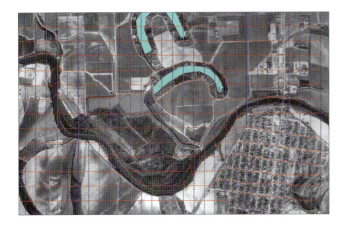

Fig. 9.5b. Preplot example.

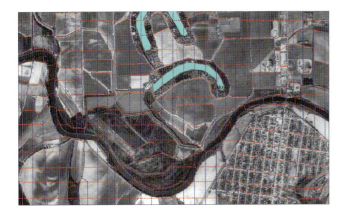

Fig. 9.5c. Typical final plan.

Table 9.1. Sample distance requirements table (Canada/*US*).

Obstruction	Nonexplosive		Explosive	
Building	50 m	*330 ft*	180 m	*300 ft*
Water well	100 m	*330 ft*	180 m	*300 ft*
Low-pressure pipeline	3 m		3 m	
High-pressure pipeline	15 m	*300 ft*	32 m for $\leq$ 2kg	$\geq$ *200 ft* (≤ *5 lbs*)
Oil and gas well	15 m	*300 ft*	up to 180 m (depending on charge size)	

output, not only for the processor but also for regulatory bodies. When final plans for regulatory bodies are submitted, sometimes they include only approximate line locations and not specific source and receiver coordinates. Final plans should also include information such as access routes and detours that the field crew can use.

Another technique for handling offset problems is called "design on the fly." When using this technique, the designer works in the field with the permit agents, surveyors, seismic layout crew, and appropriate computer resources to adjust for offsets and skips "on the fly." This technique works best when the designer has premoved many of the points and compensated with other moves and skids. With all of these premoved points loaded into the GPS units and a field design package, the designer works in the field with a radio and a computer. When a point needs to be offset, the survey crew or the permit crew calls the designer who then models the offset and compensates by moving other points or offers other possible moves. These other moves are tested in the field as the layout progresses. This technique is most effective when there is a good working relationship among all of the different groups and an excellent radio network is established.

Map coordinates are usually given using a Universal Transverse Mercator (UTM) projection. The central meridian that is used as a reference is an extremely important piece of information. Computer programs for conversion from one geographic coordinate system to another are readily available. Knowledge of the spheroid, ellipsoid, and details of the projection are of particular importance for international work. For example, in Argentina a modified UTM grid is used with an origin in the southeast Pacific. In Africa, a different spheroid affects the latitude and longitude conversion.

9.2 SCRIPT FILES

Modern field systems are controlled by script files that define how a live patch of receiver stations moves throughout the entire 3-D program as the source point sequence progresses. For small 3-D surveys where the entire 3-D is laid out and is live before taking the first source point, the sequence of source points is immaterial. However, on larger 3-D surveys, the progression of source points is of utmost importance. Time is crucial and reoccupying either source or receiver stations has to be minimized. On larger 3-D surveys, it is common that the number of available channels (geophones, cables, and boxes) is at least twice the number of channels in a live patch. This spread capability allows for more effective crew operation and patch movement. If the number of available channels approaches the size of the active patch, the efficiency of the crew plummets because the acquisition crew starts waiting on the layout crew, and standby costs become excessive. Managing the available channels and the efficiency of the layout crew can save significant amounts of money.

Most computer programs used for designing the 3-D geometry can generate script files. These files can then be loaded into the field recorder prior to shooting. The operator in the recording truck may have limited ability to alter script files. One should never assume that the operator knows how the designer anticipates the 3-D survey to be acquired as far as the source point sequence is concerned. Script files should be tested with the contractor in advance. There are enough idiosyncrasies in the file for mistakes to cause serious delays in the acquisition, and these potential problems should be eliminated before the crew is ready to start shooting. A meeting between the party chief, observer, and the designer before the crew goes to the field can identify ways to optimize field operations and reduce costs. A typical script file is shown in Table 9.2. Script files are identified by serial numbers or source-point numbers, and then the template is set. The script file does not have to be ordered sequentially. Several script file formats are available for acquisition systems such as ARAM, Fairfield, I/O, and Sercel.

Shell Oil submitted an alternate format for script files to the SEG for use in the field. This format is referred to as the Shell Processing Support (SPS) format. The SPS files should contain everything that one needs to know about a 3-D survey. The seismic processor then knows all the acquisition information required to process the data. SPS files contain four groups of files:

Table 9.2. Script file example.

This is the beginning of an I/O Script file, converted to ASCII, created from FD5.0.
The survey is 3200 m × 3200 m, with 50 m × 50 m bins
RLI = 200 m (N/S); SLI = 400 m (E/W); 297 sources 561 receivers; no roll-on; I/O numbering option chosen.
HEADER[34bytes]:Script file, software rev: 2.62
[2 bytes]10 10
Separator:4660 nscripts:297

Serial:101 Stn type:1 Src Type:1
Shot Point: LINE: 1.0 STATION: 0.0
RECEIVER PATCH: Lowest Line :1 Lowest Stn : 0
   Highest Line:16 Highest Stn: 31
ACTIVE LINES: 16
Line: 1 First Stn: 0 Last Stn: 31
Line: 2 First Stn: 0 Last Stn: 31
Line: 3 First Stn: 0 Last Stn: 31
Line: 4 First Stn: 0 Last Stn: 31
Line: 5 First Stn: 0 Last Stn: 31
Line: 6 First Stn: 0 Last Stn: 31
Line: 7 First Stn: 0 Last Stn: 31
Line: 8 First Stn: 0 Last Stn: 31
Line: 9 First Stn: 0 Last Stn: 31
Line: 10 First Stn: 0 Last Stn: 31
Line: 11 First Stn: 0 Last Stn: 31
Line: 12 First Stn: 0 Last Stn: 31
Line: 13 First Stn: 0 Last Stn: 31
Line: 14 First Stn: 0 Last Stn: 31
Line: 15 First Stn: 0 Last Stn: 31
Line: 16 First Stn: 0 Last Stn: 31

Serial:102 Stn type:1 Src Type:1
Shotpoint: LINE: 1.50 STATION: 0.0   - Different source point
RECEIVER PATCH: Lowest Line :1 Lowest Stn : 0
                                               - Same receiver patch
   Highest Line:16 Highest Stn: 31
ACTIVE LINES: 16
Line: 1 First Stn: 0 Last Stn: 31
Line: 2 First Stn: 0 Last Stn: 31
Line: 3 First Stn: 0 Last Stn: 31
Line: 4 First Stn: 0 Last Stn: 31
Line: 5 First Stn: 0 Last Stn: 31
Line: 6 First Stn: 0 Last Stn: 31
Line: 7 First Stn: 0 Last Stn: 31
Line: 8 First Stn: 0 Last Stn: 31
Line: 9 First Stn: 0 Last Stn: 31
Line: 10 First Stn: 0 Last Stn: 31
Line: 11 First Stn: 0 Last Stn: 31
Line: 12 First Stn: 0 Last Stn: 31
Line: 13 First Stn: 0 Last Stn: 31
Line: 14 First Stn: 0 Last Stn: 31
Line: 15 First Stn: 0 Last Stn: 31
Line: 16 First Stn: 0 Last Stn: 31

Table 9.3. SPS script file example.

Typical SPS files. Four files are shown on the following pages. This first file is the header file.

H HEADER FILE

H00 SPS format version num.	SPS001,07.02.95
H01 Description of survey area	ngal,,N/A,N/A
H02 Date of survey	07.02.95,07.02.95
H021 Post-plot date of issue	07.02.95
H022 Tape/disk identifier	N/A
H03 Client	N/A
H04 Geophysical contractor	N/A
H05 Positioning contractor	N/A
H06 Pos. proc. contractor	N/A
H07 Field computer system(s)	GMG/SIS,MESA,Version 1.2
H08 Coordinate location	N/A
H09 Offset from coord. location	N/A
H10 Clock time w.r.t GMT	N/A
H12 Geodetic datum,-spheroid	N/A
H14 Geodetic datum parameters	N/A
H17 Vertical Datum description	N/A
H18 Projection type	N/A
H19 Projection zone	N/A,N/A
H20 Description of grid units	AMERICAN FEET
H201 Factor to meter	0.30480061
H220 Long. of central meridian	N/A
H231 Grid origin	N/A
H232 Grid coord. at origin	N/A
H241 Scale factor	N/A
H242 Lat., Long.- scale factor	N/A
H30 Project code and description	N/A,N/A,N/A
H31 Line number format	N/A
H400 Type,Model,Polarity	1,N/A,N/A,N/A
H401 Crew name, Comment	1,N/A
H402 Sample rate, Record Len.	1,0.000000,N/A
H403 Number of channels	1,320
H404 Tape type, format, density	1,N/A,N/A,N/A
H405 Filter_alias Hz,dB pnt,slope1,	N/A,N/A,N/A
H406 Filter_notch Hz,-3-dB points	1,N/A
H407 Filter_low Hz, dB pnt slope	1,N/A
H408 Time delay FTB-SOD app Y/N	1,N/A
H409 Multi component recording	1,N/A
H410 Aux. channel 1 contents	1,N/A
H411 Aux. channel 2 contents	1,N/A
H412 Aux. channel 3 contents	1,N/A
H413 Aux. channel 4 contents	1,N/A
H600 Type,Model,Polarity	G1,N/A,N/A,N/A
H601 Damp coeff,natural freq.	G1,N/A,N/A
H602 Nunits, len(X),width(Y)	G1,N/A,N/A,N/A
H603 Unit spacing X,Y	G1,N/A,N/A
H700 Type,Model,Polarity	E1,N/A,N/A,N/A
H701 Size,vert. stk fold	E1,N/A
H702 Nunits, len(X),width(Y)	E1,N/A,N/A,N/A
H703 Unit spacing X,Y	E1,N/A,N/A
H711 Nom. shot depth,charge len.	E1,N/A,N/A
H712 Nom. soil,drill method	E1,N/A,N/A
H713 Weathering thickness	E1,N/A
H990 R,S,X file quality control	07.02.95,N/A,N/A
H991 Coord. status final/prov	N/A,07.02.95,N/A,N/A

Table 9.3. SPS script file example (continued).

S SOURCE POINT FILE

S1	10011E1	0 0.0	0 0 0.0	1010.7	25.1	0.0	1235959
S1	10021E1	0 0.0	0 0 0.0	989.8	44.5	0.0	1235959
S1	10031E1	0 0.0	0 0 0.0	971.6	61.3	0.0	1235959
S1	10041E1	0 0.0	0 0 0.0	953.0	76.3	0.0	1235959
S1	10051E1	0 0.0	0 0 0.0	934.8	95.8	0.0	1235959
S1	10061E1	0 0.0	0 0 0.0	919.5	117.9	0.0	1235959
S1	10071E1	0 0.0	0 0 0.0	903.6	136.5	0.0	1235959
S1	10081E1	0 0.0	0 0 0.0	886.5	149.7	0.0	1235959
S1	10091E1	0 0.0	0 0 0.0	860.8	161.1	0.0	1235959
S1	10101E1	0 0.0	0 0 0.0	843.1	178.8	0.0	1235959
S1	10111E1	0 0.0	0 0 0.0	825.5	196.5	0.0	1235959
S1	10121E1	0 0.0	0 0 0.0	807.8	214.2	0.0	1235959
S1	10131E1	0 0.0	0 0 0.0	790.1	231.8	0.0	1235959
S1	10141E1	0 0.0	0 0 0.0	772.5	249.5	0.0	1235959
S1	10151E1	0 0.0	0 0 0.0	754.8	267.2	0.0	1235959
S1	10161E1	0 0.0	0 0 0.0	737.1	284.9	0.0	1235959
S1	10171E1	0 0.0	0 0 0.0	719.4	302.5	0.0	1235959
S1	10181E1	0 0.0	0 0 0.0	701.7	320.2	0.0	1235959

R RECEIVER LOCATION FILE

R1	10051G1	0 0.0	0 0 0.0	1073.0	72.7	0.0	1235959
R1	10061G1	0 0.0	0 0 0.0	1090.7	90.4	0.0	1235959
R1	10071G1	0 0.0	0 0 0.0	1108.3	108.1	0.0	1235959
R1	10081G1	0 0.0	0 0 0.0	1126.0	125.7	0.0	1235959
R1	10091G1	0 0.0	0 0 0.0	1143.7	143.4	0.0	1235959
R1	10101G1	0 0.0	0 0 0.0	1161.3	161.1	0.0	1235959
R1	10111G1	0 0.0	0 0 0.0	1179.0	178.8	0.0	1235959
R1	10121G1	0 0.0	0 0 0.0	1196.7	196.4	0.0	1235959
R1	10131G1	0 0.0	0 0 0.0	1214.4	214.1	0.0	1235959
R1	10141G1	0 0.0	0 0 0.0	1232.0	231.8	0.0	1235959
R1	10151G1	0 0.0	0 0 0.0	1249.8	249.5	0.0	1235959
R1	10161G1	0 0.0	0 0 0.0	1267.4	267.2	0.0	1235959
R1	10171G1	0 0.0	0 0 0.0	1285.0	284.8	0.0	1235959
R1	10181G1	0 0.0	0 0 0.0	1302.8	302.5	0.0	1235959
R1	10191G1	0 0.0	0 0 0.0	1320.5	320.2	0.0	1235959

X RELATIONSHIP FILE

X0	0111	10241	1	1715	5001	50171
X0	0111	10241	18	3416	6001	60171
X0	0111	10241	35	5117	7001	70171
X0	0111	10241	52	6818	8001	80171
X0	0111	10251	1	1716	6001	60171
X0	0111	10251	18	3417	7001	70171
X0	0111	10251	35	5118	8001	80171
X0	0111	10251	52	6819	9001	90171
X0	0111	10261	1	1716	6001	60171
X0	0111	10261	18	3417	7001	70171
X0	0111	10261	35	5118	8001	80171
X0	0111	10261	52	6819	9001	90171
X0	0111	10271	1	1716	6001	60171
X0	0111	10271	18	3417	7001	70171
X0	0111	10271	35	5118	8001	80171
X0	0111	10271	52	6819	9001	90171
X0	0111	10281	1	1716	6001	60171
X0	0111	10281	18	3417	7001	70171
X0	0111	10281	35	5118	8001	80171
X0	0111	10281	52	6819	9001	90171
X0	0111	10291	1	1717	7001	70171
X0	0111	10291	18	3418	8001	80171
X0	0111	10291	35	5119	9001	90171
X0	0111	10291	52	68110	10001	100171
X0	0111	10301	1	1717	7001	70171
X0	0111	10301	18	3418	8001	80171
X0	0111	10301	35	5119	9001	90171

H header file
S source point file
R receiver file
X relationship file

contains general recording information, equivalent to SEG-P1 for source point locations, essentially identical to the SEG-P1 survey format for receiver location, and cross-referencing the source points and receiver stations.

SEG has adopted the SPS format as an SEG standard, and its use is widespread. Sercel 388 and Fairfield BOX instruments can read these SPS files directly to define receiver locations for each source point. Table 9.3 summarizes an SPS script file.

9.3 TEMPLATES

It is important to minimize the number of template positions in a 3-D survey. Templates describe a particular location of a recording patch within the survey and the source points associated with that patch. Moving the patch takes time, particularly when there is a limited number of channels available to the crew. Patch moves are normally accomplished through the use of roll-along switches in the recording truck. Therefore, in this chapter the term roll is synonymous with patch moves. The number of templates in a survey is greatly influenced by the decision whether or not to roll-on/off stations or lines.

9.4 ROLL-ON/OFF

Roll-on and roll-off refers to the procedure of recording with a partial patch near the edges of a 3-D survey. Generally, the crew will be able to start faster if they can start shooting with a quarter (or half) patch near the edges. Waiting for the entire patch to be laid out is time consuming and, more often than not, a waste of far offsets. When a seismic crew starts up, they generally lay out cable until there is sufficient receiver coverage to start recording the first source point. It can take about two hours to roll one line of 100 receivers, or about a day and a half to lay out 1000 channels in good terrain. Thereafter, the receiver cable is moved as the shooting progresses.

Figure 9.6a shows the fold distribution that results with the more economical roll-on and roll-off using a patch size of 10 lines of 48 channels each. Figure 9.6b shows the higher fold distribution that occurs without roll-on and roll-off. In this case, the entire survey was live. Note that the additional fold is gained mainly

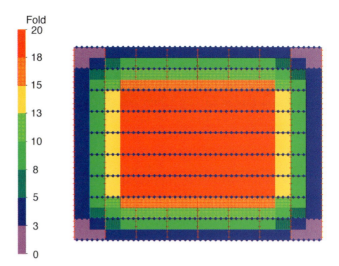

Fig. 9.6a. Fold distribution for a small 3-D layout using roll-on/off.

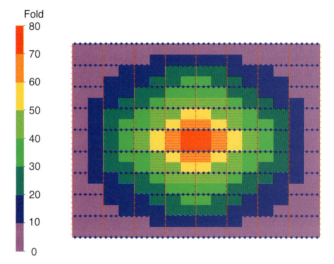

Fig. 9.6b. Fold distribution for a small 3-D layout without roll/on off.

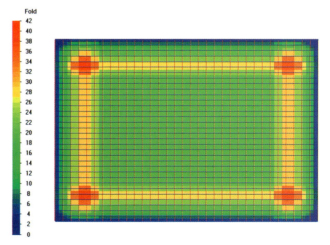

Fig. 9.6c. Fold distribution for a large 3-D survey without roll-on/off.

through far-offsets, which may not contribute to the final stack. If the cross-line dimension is much smaller than the in-line dimension, one may want to leave one or two additional receiver lines live when rolling on/off. Figure 9.6c shows the high-fold rim that develops in a larger survey when not using roll-on/off (using the same 10 lines of 48 channel patch). Applying an appropriate mute will reduce the fold in the corners significantly.

Assuming roll-on/off in the in-line direction, the number of template positions is calculated as follows (see Figure 9.6a):

$$\text{in-line templates} = \frac{\text{in-line survey size}}{\text{source line interval}} + 1$$
$$= \text{number of source lines.} \quad (9.1)$$

Example: $\frac{2400 \text{ m}}{300 \text{ m}} + 1 = 9$ in-line templates.

Assuming roll-on/off in the cross-line direction, the number of template positions is (see Figures 9.6a and 9.7):

$$\text{cross-line templates} = \frac{\text{cross-line survey size}}{\text{receiver line interval}}$$
$$= \text{number of receiver lines} - 1.$$
$$(9.2)$$

Example: $\frac{1800 \text{ m}}{200 \text{ m}} = 9$ cross-line templates.

The preceding equation is based on acquiring source points over only one receiver line interval in the center of the patch. It is noteworthy that the number of templates is independent of the patch size when rolling on/off lines and stations. The total number of templates is simply the product of the in-line and cross-line templates:

total number of templates
$$= \text{in-line templates} \times \text{cross-line templates.} \quad (9.3)$$

Example: $9 \times 9 = 81$ templates.

Hence with a swath width of one, 81 templates would be required. The total number of rolls is simply:

$$\text{total number of rolls} = \text{templates} - 1. \quad (9.4)$$

Example: $81 - 1 = 80$ rolls.

If one can traverse source points over more than one receiver line interval, the number of cross-line rolls can be reduced. This field procedure is worth considering, especially when the number of receiver lines is ten or more (see Section 9.5). If for example, a six-line patch is employed, one can roll over three line intervals (swath width = 3). The formula for the in-line templates remains as in equation (9.1), while the cross-line template formula changes as follows (assuming roll-on/off):

cross-line templates
$$= \frac{\text{cross-line survey size}}{\text{receiver line interval} \times \text{swath width}}. \quad (9.5)$$

Example: $\frac{1800 \text{ m}}{200 \text{ m} \times 3} = 3$ cross-line templates.

By increasing the swath width to 3, the number of required templates would be reduced from 81 to 27.

9.5 NO ROLL-ON/OFF

If the patch is equal to or greater than the survey size, then there is only one template when roll-on/off is not used. In the example of Figure 9.6b, there would only be one template for a patch of 10×48 channels. However if the patch is reduced to 8×24 then the number of templates increases. Assuming no roll-on/off in the in-line direction, the number of template positions is calculated as follows (see Figures 9.6b):

in-line templates
$$= \frac{\text{in-line survey size} - \text{in-line patch size}}{\text{source line interval}} + 1.$$
$$(9.6)$$

Example:
$$\frac{2400 \text{ m} - 1200 \text{ m}}{300 \text{ m}} + 1 = 5 \text{ in-line templates.}$$

Assuming no roll-on/off in the cross-line direction, the number of template positions is:

cross-line templates
$$= \frac{\text{cross-line survey size} - \text{cross-line patch size}}{\text{receiver line interval}} + 2$$
$$(9.7)$$

9.6 Swath Width 143

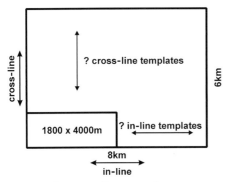

Survey Parameters

Survey Size	=	6 x 8 km
Patch Size	=	1.8 x 4 km
	=	6 x 80 channels
RI	=	50 m
SI	=	50 m
RLI	=	300 m
SLI	=	400 m

	roll-on/off	no roll-on/off
in-line templates		
cross-line templates		
total templates		

Fig. 9.7. Patch moves.

Example:

$$\frac{1800 \text{ m} - 1600 \text{ m}}{200 \text{ m}} + 2 = 3 \text{ cross-line templates.}$$

The total number of templates for a 8 × 24 patch is the product of 5 in-line templates × 3 cross-line templates, i.e., 15 templates in total.

Figure 9.7 shows a further example of a 3-D survey. The number of templates assuming roll-on/off (or not) for a swath width of one is presented as a problem to be solved for interested readers.

9.6 SWATH WIDTH

The designer has to decide if the template should include source points over only one receiver line interval, or over two, three or more, when the patch is in the middle of the 3-D survey. This technique is called a "Multiline Roll." The effect on crew speed is different depending on the approach taken. Time is the main constraint when operating a 3-D crew. It is most important to minimize the waiting time for movement of recording equipment and source equipment (drilling holes or vibrator detours), and to maximize the recording time.

Increasing the swath width decreases the amount of overlap with neighboring midpoint areas (see Section 5.2). This may have a detrimental effect on the fold, static coupling, offset, and azimuth distributions. The larger the swath width, the narrower the azimuth distribution. The far offsets increase also, leading to a larger X_{max} although the physical patch size has not changed. Figure 9.8 indicates how the areas of midpoint coverage change if the swath width is increased from 1 (a) to 3 (b). X_{max} can be calculated easily if the cross-line dimension of the patch is taken as twice the cross-line width of the midpoint coverage.

The possible change in the fold distribution is the major point of consideration. Generally the patch is rolling but for a stationary 12 × 96 channel patch. Figure 9.9 shows the fold distributions with the swath width as the only variable (the template is indicated

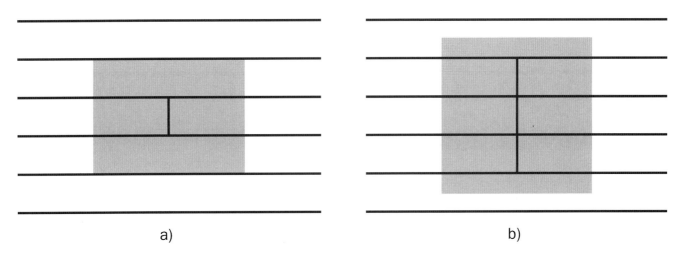

Fig. 9.8. Midpoint coverage for different swath widths (6-line patch): a. swath width 1; b. swath width 3

144 *Practical Field Considerations*

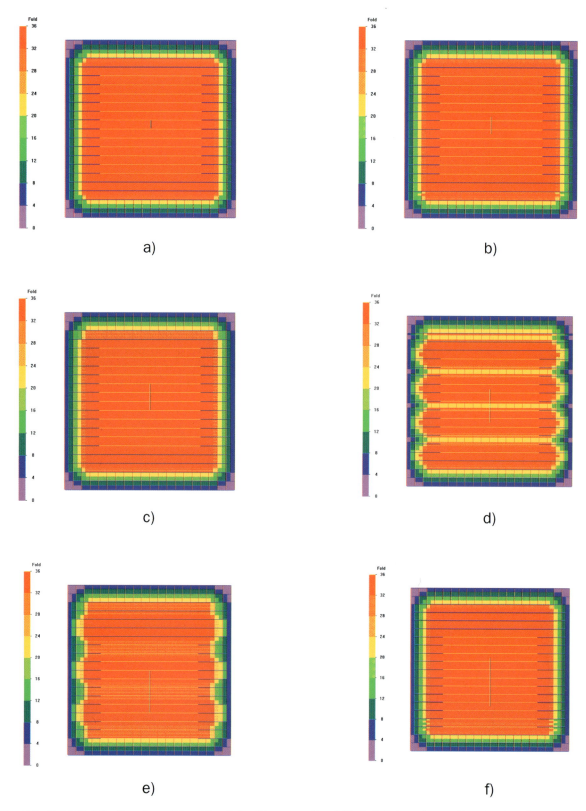

Fig. 9.9. Fold distribution for a 12 × 96 patch with: a. swath width 1, 441 templates, X_{max} 3359 m; b. swath width 2, 231 templates, X_{max} 3503 m; c. swath width 3, 147 templates, X_{max} 3653 m; d. swath width 4, 126 templates, X_{max} 3807 m; e. swath width 5, 84 templates, X_{max} 3965 m; f. swath width 6, 84 templates, X_{max} 4127 m.

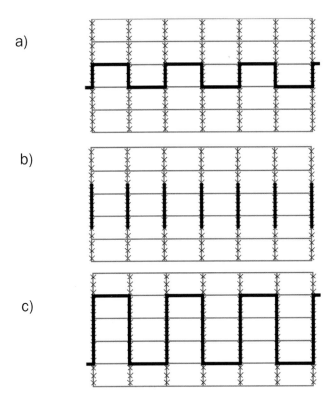

Fig. 9.10. Source point range in the middle of a patch (snaking). Receiver lines east-west, source lines north-south.

as well). The aspect ratio for this patch is 1.0 with a swath width of 1. Figure 9.9a shows that the fold in the center of the patch reaches an even 36 (with roll-on/off). This swath width indicates that the source points are merely between the center two receiver lines (compare Figure 9.10a). Figure 9.9b shows the same fold for a swath width of 2. Notice the decrease from 420 templates to 231 templates, which will have a significant impact on crew progress. The source points must be located as indicated in Figure 9.10b, which may be feasible only for dynamite crews. It would be more efficient to extend the source points to the next receiver lines as in Figure 9.10c. The fold distribution remains even (Figure 9.9c) for this 12-line patch. Some of the remaining figures show the fold striping, which can be introduced by a poor choice of swath width of 4 or 5 for this 12-line patch (Figures 9.9d and 9.9e). A swath width of 6 results in a smooth fold distribution again (Figure 9.9f). The reduction of the number of templates and the increase in X_{max} give an indication of the effects of the choice of the swath width.

Table 9.4. Swath width options.

NRL	Possible swath widths
4	1, 2
6	1, 3
8	1, 2, 4
10	1, 5
12	1, 2, 3, 6
14	1, 7
16	1, 2, 4, 8
18	1, 3, 9
20	1, 2, 5, 10

The swath width cannot be set to an arbitrary number. The following formula makes its choice dependent upon the number of receiver lines in the patch:

$$\text{swath width} = \frac{\text{number of receiver lines in the patch}}{2 \times n},$$
(9.8)

where n is an integer.

Example: swath width = 1, 2, 3, and 6 for a patch with 12 receiver lines

The maximum swath width equals half the number of receiver lines in the patch as long as source points are not duplicated. The possible swath widths are given in terms of receiver line intervals (Table 9.4).

9.7 SHOOTING STRATEGY

Always check topography maps or air photos to gain an understanding of probable access problems before deciding on a shooting strategy. An advance man will be able to identify most operational problems.

When considering the size of the patch, it is necessary to keep the movement of equipment in mind. An unnecessarily large patch that needs to be moved frequently adds a tremendous amount of time to 3-D data recording. Down times need to be balanced. For example, if recording is not possible at night, perhaps moving the recording equipment is possible. Any reduction in recording time can reduce costs, which will make the survey less expensive for the client.

Snaking is most efficient when using one vibrator crew. Snaking over three lines at a time (Figure 9.10c) is far more efficient than snaking over one line (Figure 9.10a) because 9 source points per detour versus 3 source points per detour will save a significant amount of time.

The direction in which receiver cables are laid out can dramatically affect the logistics of a survey. If possible, a crew should have enough channels to lay out receiver lines all the way across one dimension of the survey. Entire lines can then be rolled; there is no loss of efficiency in occupying stations twice or more. In long skinny surveys, the equipment layout usually depends on whether the source is vibrator or dynamite. For vibroseis, the shooting is time-consuming, and it is usually better to keep the vibrators moving while the geophone (line) crew waits. The reverse is true for dynamite. Source points can be taken quickly, so one must optimize the equipment (geophones) moves. The line crew should always travel by the shortest possible route. It is important to note that a particular design might be easy to implement using vibrators, but not be efficient for dynamite acquisition and vice versa. Therefore, there may be more than one good design to meet a particular geophysical objective.

9.7.1 Vibrator

Suppose the long dimension of the 3-D survey is north-south (Figure 9.11). For vibroseis source operation, lay out the cable north-south so each line is partially laid out and large enough for each patch, and then shoot across the lines using electronic roll-along switches. Once the vibrators have progressed through half the patch, remove short portions of receiver lines from the top of the survey and place them at the bottom.

9.7.2 Dynamite

For dynamite source operations, the receiver layout is assumed to be east-west across the short dimension of the survey (Figure 9.11), and shots are taken until it is time to move an entire line. This procedure is faster for the line crew than moving pieces of many lines. It makes no difference with regard to fold, offsets, and azimuths if the source and receiver line directions are switched.

9.8 LARGE SURVEYS

When contemplating large surveys, one has to plan how the seismic crew is allowed to record the data. A larger survey may require so many receiver stations along the receiver line direction that the crew does not have enough equipment for the number of lines necessary for the live patch. One option is to shoot in zippers;

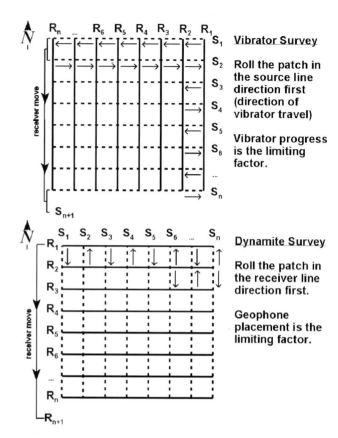

Fig. 9.11. Typical layout for a vibroseis or dynamite survey.

the other is to rent more equipment. In general, unless the amount of rental equipment already in use is excessive, zippering or redesigning the survey is the less expensive option.

Consider a receiver patch of 8 lines of 100 stations each. Furthermore, assume that the receiver lines are 250 stations long for the entire survey. If the crew has only 1000 channels available, one cannot plan a patch of 8 lines of 250 channels each. Essentially, the fold drop of the first segment needs to be overlapped with the fold build up of the second segment (Figure 9.12). The minimal amount of overlap is shown in the bottom part of Figure 9.12. Ideally, the full patch needs to be available for both segments along the common source line, or fold will be reduced and statics coupling will be affected adversely. This procedure requires reoccupying stations, which is time consuming and costly. One has to decide whether the overlap along the zipper can be reduced and to what degree.

Often there is a choice of reoccupying source or receiver stations. With vibrators, one may want to repeat source points without repeating source-to-receiver raypaths. For dynamite acquisition, laying receivers a second time may be a better alternative, al-

9.8 *Large Surveys* 147

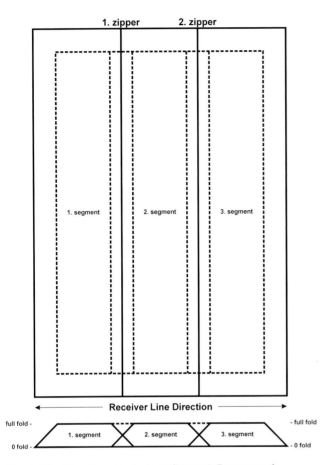

Fig. 9.12. Shooting in zippers (large 3-D surveys).

Source points for a particular template are both inside and outside the patch. From the center of the patch, the source locations extend the following distance away from the central source point:

number of receiver lines × receiver line interval. (9.10)

Therefore, the number of source points that can be taken for one template position of the full swath roll is:

number of source points per template

$$= \frac{2\,(NRL)\,(RLI)}{SI}. \quad (9.11)$$

An example of a first template A is shown at the top of Figure 9.13. After taking the source points associated with this template, the receiver patch is advanced in the in-line direction to the end of the survey as indicated. The entire patch is then rolled a full swath roll in the cross-line direction as indicated by the middle template position B in Figure 9.13. In-line rolls are repeated in the opposite direction to the other end of the survey. The template then advances by another full swath roll to the bottom position C in Figure 9.13. The bottom source point of the A template

though undesirable. The effectiveness and speed of shooting in zippers depends very much on the swath width (see Section 9.5).

The full swath roll constitutes an extremely efficient method of template advancement when the 3-D survey is very large and source points are relatively inexpensive. Source points are occupied twice and all receiver lines are rolled at once in the cross-line direction. This method allows the contractor to acquire very large 3-D surveys with only a limited channel capacity. Note that there are no common receivers from one swath to the next. Thus the statics will be completely indeterminate from one swath to the next. This method of shooting is not recommended if significant statics are expected.

Doubling the source locations increases the cross-line fold and makes it equal to the number of receiver lines active in the patch instead of half the number of receiver lines as in other geometries (see Section 2.5):

cross-line fold = number of receiver lines. (9.9)

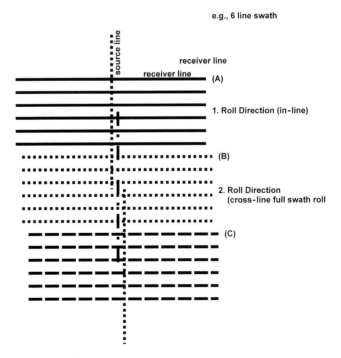

Fig. 9.13. Full-swath roll.

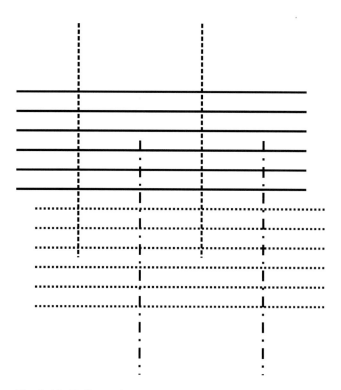

Fig. 9.14. Full-swath roll, interleaved.

position abuts the top source point of the C template position. The source points of the middle template B are completely overlapped, resulting in the increased cross-line fold. Although source locations are occupied twice, no source-to-receiver raypaths are duplicated. In practice, the full swath roll is acquired as described (often with the source lines extended even further) but is processed as a series of overlapping symmetric cross-spreads (Vermeer, 1999a). Thus not all shots that were taken into a given receiver spread are used. The redundancy in data collection (and its discarding by the processor) is more than compensated by the high rate of acquisition. In processing, therefore, the minimal data sets are simply overlapping cross-spreads (see Chapter 5), and a well-sampled 3-D is created (with the exception of the uncoupled statics described above).

If the minimum offset requirement can be relaxed, it might be possible to not reoccupy source positions but rather to interleave the source lines (Figure 9.14).

9.9 FIELD VISITS (QC)

Quality control is necessary in all aspects of the field operation, including the surveying, chaining, logging, placement of phones, quality of phone plants, vibrator synchronization, dynamite loading to depth, accuracy of source and receiver arrays, testing field parameters, alertness of the crew, and quality of monitors. Geophones should be planted vertical to the Earth's surface and not vertical to the local slope of uneven terrain. On a steep slope, arrays should be laid out in-line and parallel to the elevation contours or be bunched. Generally, elevation difference from the first to the last geophone of a group should not exceed 2 m (6 ft).

Bird-dog overseers can represent the client very well in the area of quality control. They usually have extensive field experience in different roles on seismic crews, having worked through numerous positions. They know the pitfalls to look for as well as knowing where a crew may wish to take short cuts under certain circumstances. The bird-dog visits the crew at regular intervals throughout the 3-D acquisition or is permanently stationed with them for immediate response to any problems that may arise.

The designer may also perform a QC role in the field. Many aspects of the design of a 3-D survey cannot be readily communicated to the acquisition crew or to the bird-dog. Some delays and expensive standby occurs because of inappropriate decisions made in the field. The designer who has a complete understanding of the project can quickly and effectively make decisions and changes in the field. Keep in mind that at current (2000) prices, an average seismic crew costs from $1 to $2 per second in the field. The delay time necessary to drive to town, telephone an offsite designer, explain the problem, develop a solution, and implement it in the field may cost tens of thousands of dollars.

The designer needs to determine how many consecutive source points may be dropped out of a source point sequence (e.g., 5 out of 80). Dropped source points should not be on adjacent source lines. This allowance should be made only for reasons of impossible access, pipeline and river crossings, buildings, etc. These missed source points should be substituted, if at all possible, with make-up shots. Generally these problems will be identified ahead of time, but not always. Changing weather conditions can make access impossible where previously, in dry weather, access to the source stations appeared reasonable. The more freedom the designer and the client can give the contractor and the bird-dog, the faster the field crew can operate. Cooperation among all involved parties is important in the efficient acquisition of a 3-D seismic survey.

9.10 OFFSETS AND SKIDS

When the theoretical source and receiver locations encounter an obstacle, the stations need to be moved to some other nearby location. It is essential to provide the surveyors with instructions regarding how the offsets (moves of source or receiver points perpendicular to the direction of the line) or the skids (moves of source or receiver points in the direction of the line) are to be performed.

Several priorities can be established (N. Cooper, 1997, personal communication). Most important is that each source point must be taken and each receiver must be located. This requirement maintains the average source and receiver point density and the average fold. Fold variations are introduced as the offsets and skids increase; however, these variations are less important than actually dropping the fold level because of missed source points or receiver locations. Priorities might be as follows:

1. Move the source and receiver locations by less than $1/2$ of the station interval in each direction—maintaining fold in each bin.
2. Occupy each source and receiver location—maintaining average fold within the survey.
3. Offset up to $1/2$ the line interval—offsetting introduces the least amount of fold striping.
4. Skid—skidding introduces more fold striping than offsetting (unless skidded by one line interval (Donze and Crews, 2000).
5. Do not reoccupy source or receiver locations—duplicate raypaths do not add any valuable information.

One can go as far as establishing a specific offset and skid table for each 3-D design. One example is given in Table 9.5. The lower the numbers, the more preferred this location is as an alternate to the source point to be moved (see boxed S). This particular table indicates the preference for offsets rather than skids. This preference decreases as the distance from the original location increases.

When there are several points to offset, it is preferable to offset them in a continuous smooth line (e.g., in an arc of a circle) rather than as large sudden changes in the amount of offsets and/or skids (Figure 9.15). This will produce a smoother change in the shot and receiver gathers in addition to the midpoint distribution in the subsurface and, hence, improve noise cancellation, produce less acquisition footprint, and enhance imaging (Vermeer, 1998a, b).

9.11 GENERAL CONSIDERATIONS

9.11.1 Imaging Area

In deciding how large an area needs to be imaged, several questions should be considered, such as: What are the interpreter's needs? Are all the wells tied that need to be considered? Is the survey big enough to collect energy from all dipping events? Are the migration apron and fold taper sufficiently large? It may be necessary to place extra source points outside the receiver area to accomplish some of these objectives. There may be some areas of the survey that cost a premium per unit area because of poor access or expensive permits. A decision must be made about whether

Table 9.5. Offset and skid chart for a particular 3-D layout.

21	20	S	18	17	16	15	14	13	12	11	10	S	10	11
20	19	S	17	16	15	14	13	12	11	10	9	S	9	10
19	18	S	16	15	14	13	12	11	10	9	8	S	8	9
R	R	R/S	R	R	R	R	R	R	R	R	R	R/S	R	R
17	16	S	14	13	12	11	10	9	8	7	6	S	6	7
13	12	S	10	9	8	7	6	5	4	3	2	S	2	3
17	16	S	14	13	12	11	10	9	8	7	6	S	6	7
18	17	S	15	14	13	12	11	10	9	8	7	S	7	8
19	18	S	16	15	14	13	12	11	10	9	8	S	8	9
20	19	S	17	16	15	14	13	12	11	10	9	S	9	10
21	20	S	18	17	16	15	14	13	12	11	10	S	10	11
R	R	R/S	R	R	R	R	R	R	R	R	R	R/S	R	R
19	18	S	16	15	14	13	12	11	10	9	8	S	8	9
15	14	S	12	11	10	9	8	7	6	5	4	S	4	5
19	18	S	16	15	14	13	12	11	10	9	8	S	8	9

R = receiver location; S = source point.

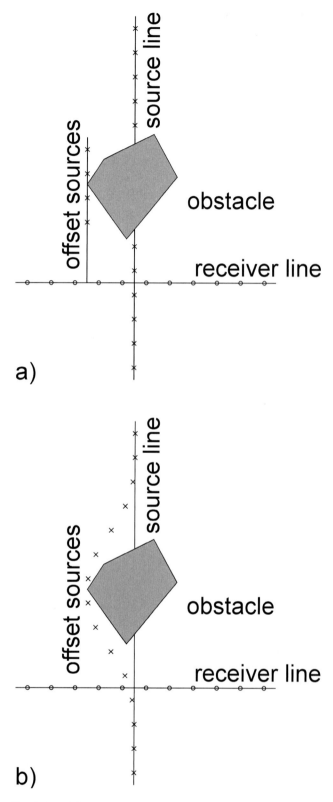

Fig. 9.15. Offset around an obstacle (after Vermeer, 1998a): a. in multiples of station intervals, and b. in a smooth line.

these areas are really needed. A re-evaluation of the non-negotiables may be called for, based on how much they cost.

It may be possible to relax the source line spacings in the aperture zone. Deep data migrate farther; therefore, a lack of short offsets may not be a problem in this zone.

Figure 9.16 indicates how one can reduce the field effort of a 3-D survey. The full-fold area in the subsurface is assumed to have a certain extent. One can either acquire the necessary data with the layout in Figure 9.16a, (bounded by source and receiver lines) or increase the receiver line lengths by one half of the source line interval on both sides and shift the source lines to the middle of the previous source line positions (Figure 9.16b), and delete one source line. Hence the total source effort is reduced. The same can be accomplished in the other direction by extending the source lines and reducing the receiver effort as indicated in Figure 9.16c. The taper zone (see Chapter 2) starts with the first midpoint coverage.

9.11.2 Cables

Certain field recording instruments have restrictions that affect the design. For example, the I/O System One and System Two each require a box every six stations. Less than six stations per box will work with I/O equipment, but to optimize use of the equipment, one should specify the number of receivers per line in the patch as some multiple of 6.

Ideally one should also have the source line interval divisible by $6 \times RI$, or $2 \times SLI$ divisible by $6 \times RI$. If the source point is not exactly between boxes, then a slight asymmetry may be introduced into the patch when using acquisition systems that cannot address receiver stations individually. The frequency of cable takeouts is impor-tant in the design. If one specifies a bin size of 35 m *(115 ft)*, hence a receiver spacing of 70 m *(230 ft)*, and the contractor has cables with takeouts every 65 m *(220 ft)*, there is a serious problem. Standard takeouts are spaced at intervals of about 65 m *(220 ft)*; although 85-m spacings are becoming more common.

9.11.3 Permitting

When cutting a standard line width of *16.5 ft*, each *half-mile* of line cut equals *1 acre* of area for crop or forestry damage calculations. In metric units, a standard 5-m line width along each 2 km of line length is 1 hectare. It is important to minimize damages by co-

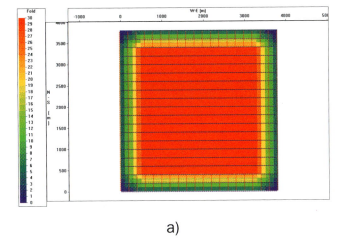

a)

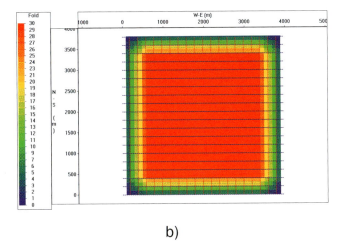

b)

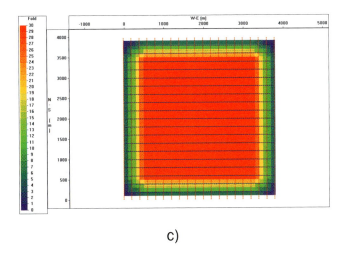

c)

Fig. 9.16 a. Maximizing the imaging area. b. Maximizing the imaging area—less one source line. c. Maximizing the imaging area—less one receiver line.

operating with landowners. Fence cuts and access routes should be used only when absolutely necessary, and one must always have prior written permission to access private lands.

Low impact seismic (LIS) guidelines were introduced in Canada in 1993. Source line widths must be kept to 5 m (6 m for vibrators) and receiver line widths to 1.5 m if hand-cut or to 3 m if cat-cut (if allowed). Generally, only trees that block the line of sight are removed. Geophones are usually carried in by hand, flown in by helicopter, or brought in on quads (ATVs). For narrow source lines, dynamite holes may need to be hand-drilled. The minimal soil disturbance of LIS programs benefits the regrowth of tree cover and impacts wildlife habitat less than standard line cutting. However, line cutting under LIS guidelines may cost as much as 30% more than standard line cutting (Wiskel, 1995).

9.11.4 Safety

The safety of the crew must be considered in the design of the survey. Subtle moves of the receiver and source points have insignificant effects on the fold and offset distribution but can make jobs much safer. For example, if the computer design puts the theoretical location on a rocky slope or a cliff, it might be best to move the point to a flatter, safer location and still keep the midpoints within the bin. Only careful modeling can determine these safety changes. Source and receiver locations can be moved for safety as well as for better geophone placement and coupling.

Daily safety meetings are necessary to make crewmembers aware of any possible hazards regarding dynamite handling and general field operations. In many countries, medical support needs to be on site, not only for the benefit of the crew, but because it is required by law. The associated cost has to be taken into account when planning the 3-D survey. An accident reporting procedure should be defined so that all necessary contractor and operator personnel and government agencies are notified in a timely fashion. The contractual agreement defines the responsibilities of each party involved in the 3-D seismic survey should any accidents occur.

A hazard registry or hazard book should be started to document the safety problems on the job and every employee should be knowledgeable in these problems. The hazard registry is an extension of good access maps that should be provided to each of the crews on layout. Also, good surveyors who document problems and safety concerns are vital to the success

152 *Practical Field Considerations*

Fig. 9.17. Safety equipment.

of your safety program. Equally important is getting this information to the crewmembers in a timely fashion.

In some cases, the designer may be the least experienced person in the field, and personal safety may be overlooked if there are no safety checks for people that are more accustomed to office work. A good travel kit (Figure 9.17) can combat these problems for fieldwork, combined with a sign-in and sign-out buddy system so that someone knows the field locations of all personnel day and night. Spending a cold night hurt in the field could be at the least inconvenient and uncomfortable and at the worst deadly.

9.12 FIELD EXAMPLES

The following figures compare the fold distribution prior to 3-D data acquisition (office design) to the actual fold distribution achieved (field reality). Figure 9.18a is the proposed layout of a 3-D survey before it was recorded. For comparison, the actual surveyed layout is shown in Figure 9.18b. Analysis of the CMP fold for different offset ranges revealed some severe inadequacies in the fold of the near traces (e.g., 0-800 m; Figures 9.19a and 9.19b), in particular in the center of the survey near a river. In addition, some east-west fold striping is introduced due to the offsets and skids. Extra source points may need to be added in various places to solve these problems and to compensate for real-life situations.

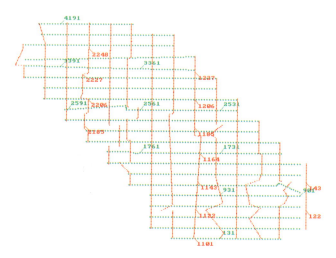

Fig. 9.18a. Layout of a survey made in the office before going to the field.

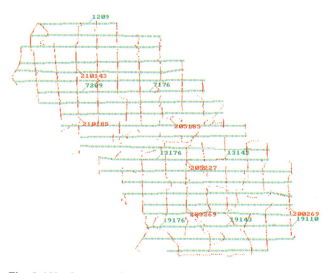

Fig. 9.18b. Layout of the survey reflecting field conditions.

9.13 Field QC (Data)   153

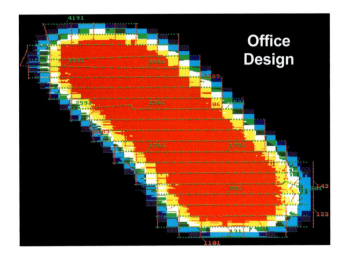

Fig. 9.19a. Office design with short offsets.

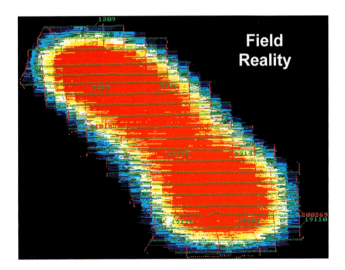

Fig. 9.19b. Field reality with short offsets.

9.13 FIELD QC (DATA)

In 3-D seismic surveys, the volume of data can be overwhelming. It is essential to have methods that quickly identify errors and/or bad data. Modern field QC uses seismic trace attributes (energy levels within specified windows, energy decay factors, first arrival times) in combination with survey information (shot and receiver positions) to create new quality control measurements and displays. Such QC displays can usually pinpoint location errors and other problems at both shot and receiver positions.

There are three basic reasons to process data in the field:

1. To verify accuracy of trace headers—3-D geometry.
2. To generate quasi-real-time interpretable results for modifying the field program or siting a well.
3. To enable fast-track processing in a remote data-processing center.

In-field quality control processing verifies the following:

positional data quality
seismic data quality
seismic/positional data relationship

The most important item is the seismic/positional data relationship. By examining the relationship between seismic data and the position in which the energy was initiated (source point), the subsurface it traveled through (CMP), and its recording position (field station or receiver), one can determine the accuracy of the position data and the quality of the seismic data. Those relationships help determine the first-break times, trace to trace variations in amplitude, and the expected spatial continuity of data attributes such as exponential amplitude decay.

9.13.1 Positional Data Quality

To verify position data, one can perform conventional offline geodetic QC, but the most useful method is to overlay final survey coordinates onto a map or image (TIFF or DXF files). The visual tie between shot and receiver positions and the actual ground is a compelling verification of accuracy.

9.13.2 Seismic Data Quality

A display of seismic data can reveal much about potential position errors and answer questions, such as:

Do the first breaks look as expected (in comparison to a theoretical first-break template overlay)?
Are the number of shots and number of traces per shot as expected?
Is the data quality as expected (are there any reflections)?
Is the noise level acceptable (are there a lot of bad traces/receivers)?

9.13.3 Verify Seismic/Positional Data Relationship

The relationship between shot, CMP, and receiver positions dictates how the data appear. In other words, the first breaks appear at a certain time and the amplitude of the data follows the laws of spherical divergence and inelastic attenuation. A semirigorous method includes the verification of shot, receiver, and CMP information, and the continuity of trace attributes to verify relationships (data and position).

The seismic data trace attributes may be chosen from the following:

- The first-break times are related to the horizontal distance between shot and receiver. As a result, one can verify relative position correctness.
- The average amplitude within a window around the first breaks and within the data indicates how much spherical divergence there has been. Consequently, one can exploit the relation between amplitude and shot/receiver offset. This can verify first break information in the case of bad picks or no picks.
- The amplitude decay factor indicates the path taken by the energy traveling from shot to receiver. This attribute is therefore related to some area of the subsurface. Such an attribute can be used to verify correctness of global position by studying its spatial continuity.
- The frequency at maximum amplitude is again related to the subsurface path. It is, however, most useful in detecting bad traces. Frequency is generally the same for all good traces but can be quite different on noisy traces (often associated with a receiver station).
- The signal-to-noise ratio is calculated as the ratio of rms amplitude within a window ahead of the first breaks (the "noise" window) to the rms amplitude in a window that includes signal within the data. The signal-to-noise ratio can be affected by many factors such as surface conditions, offset, ground roll and other linear shot noise, random noise, and bad channels.

Other attributes can be similarly classified according to their use in verifying or determining shot and receiver position.

Relationship Diagnostics for Shot/Receivers

In a rigorous QC procedure, a map of shots and receivers may display the minimum, maximum, and average (mean) of these items:

elevation of shots and receivers
shot and receiver line numbers

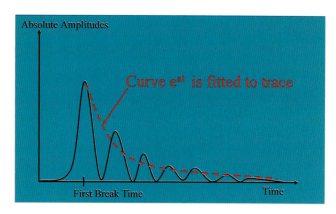

Fig. 9.20. Definition of the amplitude decay parameter.

shotpoint and field station numbers
shot, receiver, and CMP fold
depth of shot, uphole times, datum elevation
mean rms amplitudes
amplitude decay factors
maximum frequency
signal-to-noise ratio

Definition of amplitude decay:

A single parameter a can be used to define the exponential decay of a data trace by a least-squares fit of the curve e^{at} to the peaks of the absolute trace amplitudes (Figure 9.20). This parameter is directly related to the energy travel path, and therefore to subsurface geology. This parameter should show spatial variations that have a logical relationship to geology.

Definition of trace statistics:

A single parameter, "maximum frequency," can be defined for each trace. This is the frequency of the maximum amplitude of a spectrum derived from some window of the data (Figure 9.21).

Relationship Diagnostics Shots / Receivers

The following header word values for each source point might be displayed on a map of sources and receivers.

X_{mid}, Y_{mid}
shot receiver azimuth
rms amplitude
first-break times

Relationship Diagnostics Sources/Receivers:

Other relationships are examined through the use of graphs and cross-plots, such as shotpoint number,

9.13 Field QC (Data)

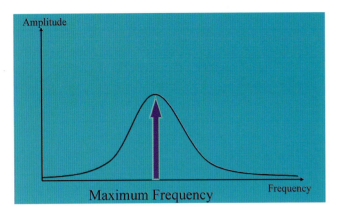

Fig. 9.21. Definition of the maximum frequency parameter.

offsets for each trace, and the rms amplitude of a window around the first breaks. An excellent agreement between the offset and amplitude is a further confirmation of relative position accuracy.

Predicting source locations from first-break times at each receiver should be plotted as a check. The source location can be predicted based on the knowledge of all receiver coordinates for each shot and the first-break times. In Figure 9.22, two small arrows on the right-hand side indicate source points that should be moved. In other words, the source point at the "tail" of the arrow should be moved to the "head" of the arrow to satisfy the combination of first-break pick times and source location. The maximum amplitude times appear to be centered around the head of

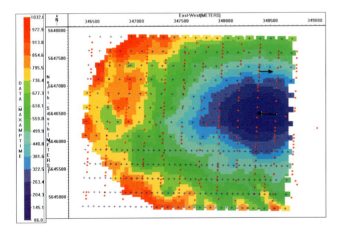

Fig. 9.22. Prediction of shot locations from first break times and verification using the times of maximum amplitude for each trace.

the arrow, which indicates the correctness of the prediction.

Predictions can be repeated based on rms amplitudes of a window over the first arrivals. In general, this is not as accurate as first-break times but can help to confirm the prediction. The receiver position can also be predicted based on the knowledge of all source coordinates (source points fired into each receiver) and the first-break time.

Other QC diagnostics:

Many QC diagnostics can further confirm (or deny) the accuracy of geometry information. Some of the most useful and enduring of these are:

- 100% displays
- field stacks
- raw record displays
- use of in-line/cross-line displays
- time slice

The main use of each is to emphasize spatial continuity of data attributes. Discontinuities generally indicate the presence of a survey or data-coordinate error.

Further QC diagnostics may also be used to check the integrity of the seismic data itself. Such diagnostics tell more about the recoverable signal content than any relationship to position information. Such diagnostics may include:

- amplitude spectra
- filter panel plots (define signal bandwidth)
- deconvolution panels
- Farr diagrams [apply largest minimum offset (LMO) to each shot and plot first-breaks in cascaded format at their surface position]
- *f-k* spectra

Field QC can establish the accuracy of the survey geometry and offers information about the signal content of the data. In general field QC can identify, quantify, and assist the following:

- position/seismic attribute relationships
- list of dead/reverse traces (shots, receivers, channels)
- data preprocessing
- velocity analysis
- processing trials (establish sequence and parameters)
- list of unusual data aspects

Field QC can enable fast-track data processing. All of these items contribute to faster throughput in the processing center and to a more successful result from the 3-D survey.

Chapter 9 Quiz

1. How does a final plan differ from a preplot?
2. Describe the essential elements of a script file.
3. Does roll-on and roll-off result in higher or lower fold?
4. When are snaking (movement of sources across multiple receiver line intervals) and ping-ponging used?

10

Processing

This chapter is not intended to be a thorough treatment of seismic data processing; the topic of data processing is thoroughly discussed by Yilmaz (1987). Rather, this chapter introduces and briefly discusses some processing issues that are closely related to 3-D acquisition.

10.1 PROCESSING

It is of utmost importance to find a processor who can efficiently conduct the required 3-D data analyses. Quality control (QC) is so important that at least one person needs to be committed to examining the progress of the processing at numerous stages throughout the 3-D project. Larger companies may have a dedicated processing QC group rather than having the interpreter be responsible for the processing QC. The best processor and the best QC person need to be brought together to achieve optimal data quality.

Field processing systems are now so small that many acquisition companies have full processing capabilities on the job site. The client has to evaluate the advantages of faster data-processing turnaround with the additional costs involved with such amenities (Duncan et al., 1996). For larger 3-D data sets, one has to make sure that the field processing center can effectively carry out the tasks. In such cases, the field processing center may be used only for quality control, data editing, or first-break picking. The quality control must be just as efficient as if the data were processed in an office environment.

Preparation of the 3-D geometry itself is by far the largest task in processing. Some processors estimate at least 50%—and some up to 70%—of the total time spent in processing concentrates on 3-D geometry. Good QC tools are essential to the verification of the 3-D geometry.

To avoid surprises, the processing stream needs to be defined and a pricing schedule established before processing commences. Most companies prefer only tape or CD-ROM copies for workstation manipulation of the data. However, at times it is necessary to perform a quick check on some aspect of the data, and, in such cases, some form of hard copy might be preferable. Selected portions of the 3-D data volume can be displayed on paper or reproducible film. Time slices of data volumes at various stages of processing are effective quality control tools. It is essential to overlay the recording geometry on these time slices to detect geometry-induced artifacts (acquisition footprints).

10.2 PROCESSING STREAM

Processing skills, experience, and the appropriateness of a given software package to process seismic data from a specific geographic area are not uniform from one processing company to another. Therefore it may be appropriate to have the 3-D data processed by several processing contractors as part of a risk reduction procedure (Bouska, 1998b).

A typical data-processing stream is shown in Table 10.1. The first group of items represents a basic processing stream for which turnkey prices are often provided. The second list of items represents possible intermediate add-on services for which unit prices can be quoted. The last three groups are minimum deliverable products (paper, film, and tape) one should expect to receive as part of the turnkey price. Some contractors may offer additional add-on services on a price-per-unit basis.

The importance of correct survey information needs to be emphasized. A change as small as 15 m (*50 ft*) in accurate line or point position can play havoc with the processing for an entire 3-D. In the absence of correct survey information, the processor can still use first arrival times to place the source points in their correct spatial position. However, this is a time-consuming task, which is best avoided.

A set of standard film displays for 3-D surveys can be helpful even when workstations are used for the interpretation. It is recommended that regional displays of every tenth line be available, as well as zone-of-interest plots for every line or every second line, to retain detailed information in hard-copy form that will be needed for picking well locations. All of these displays should be made for both polarities in both directions.

10.3 REFRACTION STATICS

Source-to-receiver times (first-break times) are essential to solve for refraction statics. If such measurements are not present in both in-line and cross-line directions there is no tie between lines or portions of lines (as is the case for narrow receiver patches).

Short offsets are needed for shallow refractors. Therefore the receiver line spacing must be small enough to adequately sample shallow refractors because a small X_{min} is necessary to define the first refractor interface. Note that all cross-line and in-line coupling is lost when the receiver line interval is greater than the offset needed to get refraction information from the shallowest refractor. This offset is shown as X_1 in Figure 10.1.

The equations used to solve for the velocity of each refractor involve source-receiver pairs and their midpoints. In any regular geometry (straight, brick, zigzag, etc.), these equations are not coupled (see Reflection Statics below). The only way to solve such uncoupled equations is to use a velocity smoother over several CMPs. This smoothing technique means that undetermined long-wavelength errors will be present in the velocities and may appear in either the source or receiver delays.

The use of a narrow receiver patch also introduces problems. It is important that the patch be wide enough to collect cross-line measurements for deep refractors. Otherwise, there are few direct measurements of deep refractors from any given source point

Table 10.1. Processing requirements.

Processing sequence (turnkey price):
    Demultiplex and edit field data
    Gain recovery
    Instrument and geophone dephasing
    Geometry and refraction statics analysis
    Deconvolution and filter tests
    Surface-consistent deconvolution and scaling
    Brute stack
    Velocity analysis
    Automatic surface consistent statics
    Intermediate stacks
    Final velocity analysis
    3-D trim stack with final statics
    FX-decon
    3-D migration after stack, one-pass
Options possible (price per unit):
    3-D interpolation
    Bin-borrowing or mixing
    Rebinning
    f-k noise attenuation
    DMO
    Multiple attenuation (e.g., Radon transform)
    Depth migration—mainly after stack
    Prestack time or depth migration
Deliverables (paper):
    Selected shot records, before and after NMO
    Common-offset stacks with mute displays
    Decon tests
    Narrow and broad-band filter tests
    Brute stack
    Velocity analysis displays (e.g., semblance)
    Intermediate stack
    Final stack
    Migration
Deliverables (film):
    Map of source and receiver locations
    Subsurface bin and coverage maps
    Migration (e.g., every tenth line, both polarities, both directions)
Deliverables (SEG-Y tape):
    FX-decon—unfiltered and filtered
    Migration—unfiltered and filtered

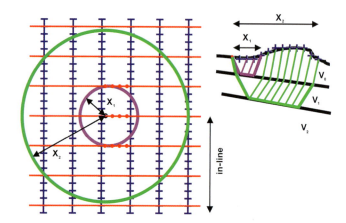

Fig. 10.1. Refraction criterion—plan view and section view showing relationship between line spacings and refraction statics. Compare with Figure 2.17b.

to its associated receivers in the cross-line direction. The variations of such refractors in the cross-line direction can be determined only by combining variations from a receiver line to an immediate neighboring receiver line. Such a combination will almost certainly be prone to error. It is therefore essential to have in-line and cross-line offsets along both source and receiver line directions that define all near-surface low-velocity layers.

10.4 VELOCITY ANALYSIS

A stacking velocity analysis should be performed as often as necessary to provide a good velocity field. At times, velocity analyses are done every 500 m in both directions to create a grid of velocity control points. On small surveys or surveys with more geological complexity, one may wish to have a velocity analysis grid point at every line intersection. Vertical traveltime calculations are based on straight raypath theory and constant velocity assumptions. Semblance analysis is used to estimate maximum coherence along a moveout curve, which then defines the stacking velocity function. Semblance analysis should not be used when the interpreter is expecting AVO effects, because one of the basic assumptions in semblance analysis is that the signal amplitude does not vary with offset.

Sometimes a super bin, i.e., several bins grouped together, is used to create a good mix of offsets. All offsets should be grouped by azimuth because velocity analysis is a directional property in a 3-D survey. Each set of offsets in each azimuth range must adequately define normal moveout (NMO) curves, which means the delta-t change should exceed one wavelength (Figure 10.2a). Care must be exercised to choose a super bin that is small enough to not smear geological structures (Figure 10.2b).

The velocity resolution of each CMP bin in a real survey can be calculated by creating a synthetic event with a specified velocity and time, and then stacking the synthetic data at different velocities. An example of such a synthetic event ray-traced for an actual offset mix across several CMPs of a real 3-D survey is shown in Figure 10.3. Several synthetic CMP gather traces before NMO are shown in Figure 10.3a. The result of stacking one of these gathers (traces belonging to one CMP) at many different velocities is shown in the bottom display (Figure 10.3b). The correct velocity is reasonably obvious (approximately CVS trace 201), but a small error in picking is still quite possible, particularly in real data.

This fact is further borne out by looking at a semblance analysis of the same data in Figure 10.4. The offsets that record the synthetic data as a flat event are not regular. As a result, the pick for the correct velocity cannot be made with total accuracy. On a computer-driven semblance plot, it was possible to pick a velocity 50 m/s either side of the correct answer and still satisfy the criterion of picking the maximum amplitude in the semblance. The picks can be made on the first peak, the trough, or the second peak. In this case, the trough is picked. In real data there is, of course, no such certainty. This velocity picking variation (± 50 m/s in this example) is referred to as the velocity resolution.

Figure 10.5 defines the velocity resolution graphically. This plot shows the maximum amplitude of the stack trace as a function of the NMO velocity that was applied before stack. At the correct NMO velocity, the amplitude of the stack trace achieves a maximum value. There is some uncertainty as to exactly where this maximum amplitude occurs. Obviously the amplitude of a stack trace is much lower when the wrong velocity is used, whether the velocity is too high or too low. The breadth of the velocity versus amplitude curve at this level (e.g., 3% off the peak) is a finite quantity in velocity units that is defined as the *velocity resolution*. In effect, velocity resolution is a single quantity that expresses the ability of the offset mix in a CMP bin to correctly focus a specified target. This velocity measure is affected by the mix of offsets and is in fact a powerful indicator of good and bad offset distribution.

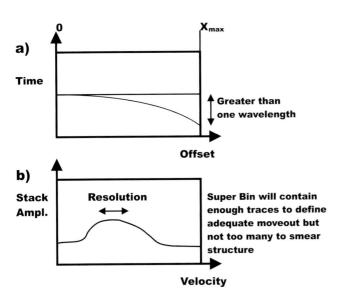

Fig. 10.2. Velocity analysis concepts for: a. NMO and b. super bins.

160 Processing

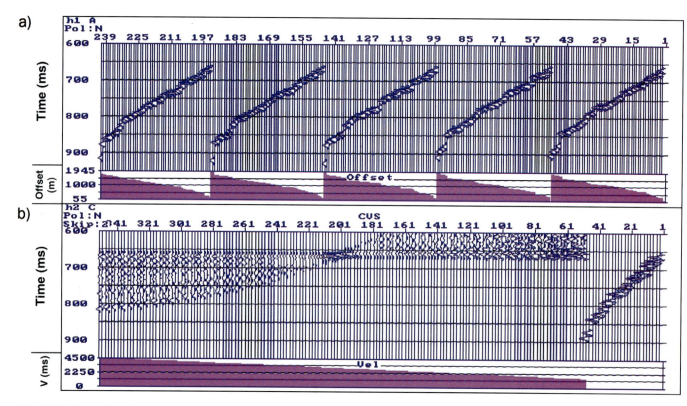

Fig. 10.3. Several gathers as a function of offset (a). One gather stacked with different velocities (CVS) (b).

This analysis was done for a proposed target and a real 3-D geometry with an rms velocity of 4000 m/s and a two-way time of 1.0 s. The breadth of the resulting stack amplitude versus velocity at the 95% level is plotted as a color value in each bin in Figure 10.6. Basically, a good mix of offsets leads to good resolution. The velocity resolution ranges from 200 to 700 m/s around the target velocity. A study of older 2-D data in the same area can reveal potential S/N to be expected, and hence the appropriate level of uncertainty (the 95% level used in this example) can be determined for a survey. Any unacceptable velocity resolution must be corrected with the addition of more (usually longer) offsets in the bad bins.

The example in Figure 10.6 is interesting because there is an obstacle (a river) crossing the source and receiver lines. The source and receiver points in the

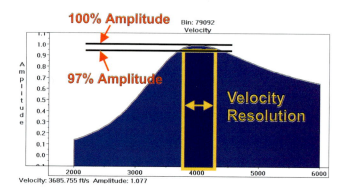

Fig. 10.4. Velocity analysis—semblance concept.

Fig. 10.5. Velocity resolution definition.

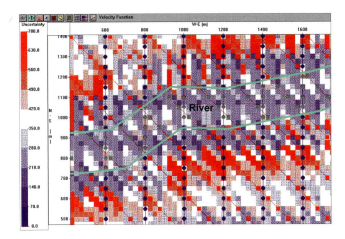

Fig. 10.6. Distribution of velocity resolution over an exclusion zone (river).

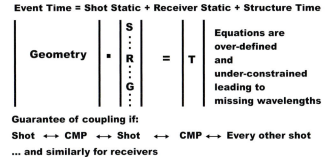

Fig. 10.7. Reflection statics equation (surface consistent).

river were excluded. Thus the offset mix in CMP bins near the river was affected. The effect of the river is mirrored in bins that are parallel to the river and positioned up to 200 m on either side. In this zone of bins, there is a higher value of velocity uncertainty, indicating that a seismic data processor must take care in selecting zones of CMPs for velocity analysis, especially near excluded areas in the 3-D survey. In the bins that fall under the location of the river, the velocity resolution is improved because there are relatively more far offsets that help improve the accuracy of the velocity determination.

10.5 REFLECTION STATICS

In the processing step of reflection statics, the traces of each, or possibly many, bins are summed to form a stable and representative pilot trace for each CMP bin. Crosscorrelation of the individual traces that enter into a stack for a particular bin with the pilot trace for that bin determines the static time shift values to be applied to each trace. Matrix calculations separate the static corrections for each source and receiver location to create what is commonly called a *surface-consistent solution*.

Statics coupling (Wiggins et al., 1976) implies that each receiver position is fired into by many source points and vice versa. Crosscorrelation methods detect time differences between adjacent receivers and adjacent source points. These time differences, when multiplied by the inverse of the geometry matrix, provide the surface-consistent static values at each receiver and source position. If the inverse of the geometry has holes in it, then the statics are undetermined for certain spatial wavelengths. The absolute static values can be thought of as being composed of a sum of different spatial wavelengths.

The occurrence of holes usually means that the geometry matrix is under-determined; that is, equations that define the statics for some receivers or source points are missing. In other words, there are too few equations to overcome potential picking errors (Figure 10.7). The number of equations required for surface-consistent static calculations is controlled by the number of source-receiver pairs that create reflection points in each bin.

Figure 10.8 illustrates the different source-receiver pairs that contribute to neighboring CMP bins. This orthogonal survey example demonstrates that if source points 1, 5, 9, ..., contribute to one CMP bin, then source points 2, 6, 10, ..., contribute to the next CMP bin, and source points 3, 7, 11, ..., to the bin after that. This means that in all the equations for each source-receiver pair, there will be no equations that link the shot static for source point 1 to that for source point 2, or the shot static for source point 3 to that for source point 4. Instead, source point 1 is

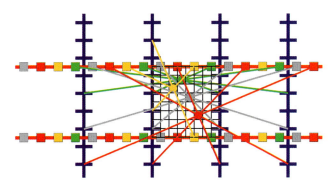

Fig. 10.8. Reflection statics—coupling.

linked to source point 5 (and 9, 13, etc.). In this case there are four families of equations that are not connected to each other. The ratios of *RLI/SI* and *SLI/RI* determine the number of families for receiver and source static solutions.

It is now known (Wisecup, 1994) that all regularly spaced straight-line geometries (brick, zig-zag, etc.) are uncoupled for statics (except the fully populated 3-D). That is, there is more than one independent solution for source and receiver statics. To tie independent solutions together, Wisecup (1994) suggests the use of smoothers in the structure term. This normally gives reasonable answers (no cross-line "jitter"), but can leave errors in the long-wavelength statics and may not adequately resolve a structural anomaly. In the Flexi-Bin® design, all sources and all receivers are coupled, because of the nonrepetitive nature of the relative positions of sources and receivers at the line intersections; this causes the interleaving of the cross-spreads (see Chapter 5).

In the limit, a 3-D survey can be shot as a series of 2-D lines. In such a case there is no coupling of solutions between adjacent lines; hence each solution can be independently high or low. As more receiver lines are added to the patch, an element of cross-line coupling is introduced, and the solutions exhibit more continuity in the cross-line direction. Note that the full swath roll (see Chapter 9) is uncoupled from swath to swath, except through the source points. Hence if there are any appreciable statics (more than 10 ms), this geometry should not be used because shot statics are not repeatable from one swath to the next.

Statics coupling can be assured by creating irregularity in the source, receiver, or line spacings (Figure 10.9). Another option is to add source lines in the receiver line direction and receiver lines in the source line direction. It should be noted that a geometry that is good for statics may not be optimum for imaging. In other words, near-surface imaging, which is what statics offer, may not be the ultimate objective, but rather a 3-D subsurface image is sought. Compromises, therefore, must always favor the subsurface targets. A computer program should be used to check each layout.

10.6 DIP MOVE-OUT

While NMO corrects for the time delay on an offset trace by moving the amplitudes to earlier times on the trace, DMO moves the data up-dip to the correct position where a zero-offset trace would record a dipping reflector (Figure 10.10). Migration then moves the energy to the correct horizontal and vertical subsurface location. Russell (1998) provides a good summary exercise for the DMO process.

DMO is a dip-specific application of prestack partial migration followed by NMO and stack. Hale (1984) discusses the process in detail. Constant-velocity DMO is the most commonly applied method. Time-variant DMO has been described by Meinardus and Schleicher (1991).

The DMO ellipse has the biggest effect on data at shallow times and at far offsets. Varying surface geometry may introduce amplitude anomalies due to different offset and azimuth mixes, that may or may

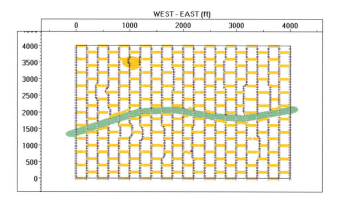

Fig. 10.9. Irregular geometry provides coupling (artificial example).

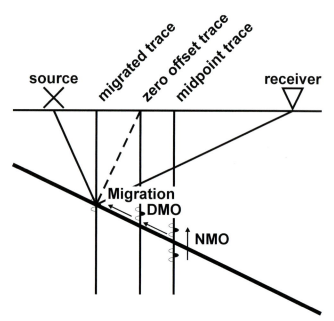

Fig. 10.10. Data movement during NMO, DMO, and migration (after Sheriff, 1991).

not be compensated for in processing. Three-dimensional DMO needs as many offsets as possible in each azimuth range to form the reflections by constructive interference. Otherwise, imperfect image reconstruction occurs. This construction of offset sections can be difficult in 3-D surveys because of midpoint scattering and source-receiver azimuth variations. Therefore, not all 3-D designs lend themselves to the successful application of DMO.

Let H be the source-receiver offset, V the average velocity, and t the time of interest. Then the energy at time t in a CMP bin is derived from all source-receiver pairs that cross the bin and whose midpoints are within $H^2 \div (2 \times V \times t)$ of the bin center as shown in Figure 10.11 (Deregowski, 1982). Fold and the number of unique offsets in each bin after DMO at time t are good measures of the DMO response. From Figure 10.11,

$$\text{DMO radius} = \frac{H^2}{2Vt}. \qquad (10.1)$$

Example: if $V = 3000$ m/s $H = 1000$ m $t = 1.0$ s

DMO radius = 170 m   (or about 7 bins if $B = 25$ m)

or   if $V = 10\,000$ ft/s $H = 3000$ ft $t = 1.0$ s

DMO radius = 450 ft   (or about 4 bins if $B = 110$ ft)

In other words, all traces that have an offset of 1000 m *(3000 ft)*, a source-receiver azimuth that passes through a central bin, and a midpoint that is within 170 m *(450 ft)* of that central bin, will contribute energy to the central bin. Note that a 5° dipping reflector in this situation has its energy shifted a horizontal distance of 15 m as defined by the equation:

$$\frac{H^2}{V \times t} \times \sin\theta \times \cos\theta, \qquad (10.2)$$

where θ is the reflector dip.

The central sample in Figure 10.12 is created from many other traces that provide contributing energy. The farther away a trace is from the center trace, the steeper the dip of the energy that contributes to the center sample. Constructive interference ensures that dips are reconstructed (imaged) at their proper zero-offset position where stacking takes place rather than at the CMP position. Migration after stack then moves the dips to their correct geologic position (CDP). If traces are missing, the constructive interference is disrupted. This effect is usually referred to as geometry imprinting (or acquisition footprint, see Section 10.8). Poor geometry design may create holes in the offset plane, and the lack of certain offsets in a regular pattern may impede the constructive interference process for certain dips. A sample at time t is reconstructed as:

sample at time $t = \Sigma$ {all offset traces [Σ all dips]}
= Dip 1 traces (Σ all offsets)
+ Dip 2 traces (Σ all offsets)
+ Dip 3 traces (Σ all offsets) + ...

Missing traces in the common-offset panel will not contribute to certain dips. The number of live traces contributing to each dip range are counted. The sum of the first and second moments of the DMO operators at each CMP position for each offset range can also be used (Jakubowicz, 1990; Beasley, 1993).

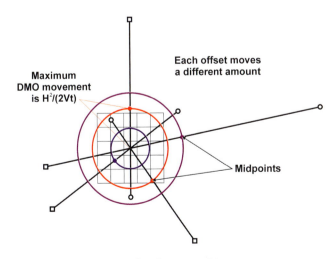

Fig. 10.11. DMO contributions to a bin.

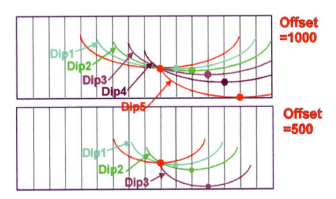

Fig. 10.12. DMO—dip decomposition.

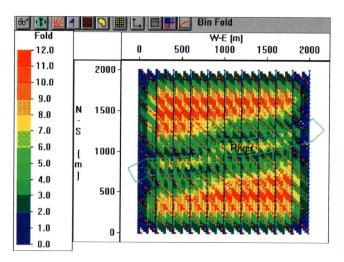
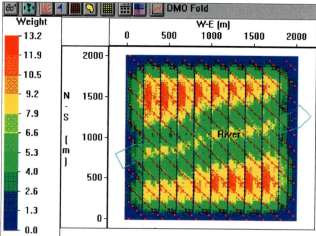

Fig. 10.13. CMP fold (left) and DMO-weighted fold (right).

A good measure of the DMO constructive interference is to calculate the weighted DMO fold in each bin for a specific target time and velocity (Crouzy and Pion, 1993a, b). This weighted DMO fold is constructed by considering all possible contributions to one output CMP bin. Energy from many source-receiver pairs having axes that cross that bin is smeared by DMO into that chosen output bin. The energy is stronger or weaker depending on how far the energy has traveled along the DMO operator ellipse. Weighting functions have been published and allow the calculation of the weighted sum of all the contributions for any given bin.

Figure 10.13 shows a comparison of midpoint fold and weighted DMO fold for an actual 3-D geometry with a river through the center of the survey. The greyed-out stations in the center indicate missing receivers and sources. The increased contribution of the far offsets to the DMO fold under the river leads to a higher DMO fold there than in the bins away from the obstacle. Such variations in weighted DMO fold can lead to anomalies in the data due to S/N variations that mirror the fold.

In addition to DMO fold anomalies, which may exhibit geometry imprinting as shown above, DMO responses for dipping events recorded with different geometries can be calculated. These responses show severe phase and amplitude changes, indicating that such dipping targets, if present on real data, may show strong geometry imprinting (Connelly and Galbraith, 1995). In the Connelly and Galbraith investigation, the target is a pseudodipping event where each CMP bin is assumed to have a specific dip at a specified time and azimuth, as depicted in Figure 10.14.

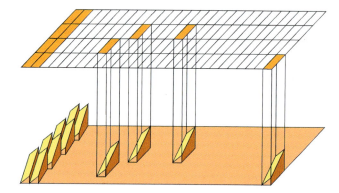

Fig. 10.14. Pseudodip in each bin for DMO model.

Calculation proceeds by adding the contributions of each source-receiver pair to the pseudodipping event at the specified target time in every bin throughout the survey. A perfect zero-phase response would indicate that the event is perfectly imaged by DMO in that bin. A less-than-perfect response indicates that there is an insufficient number of contributions to the dipping event in that bin.

Figure 10.15a shows the geometry where the algorithm was applied to a standard orthogonal geometry. Source point and receiver in-line spacing is 50 m. Source and receiver line interval is 200 m. The patch is 6 lines by 32 stations per line, giving a regular CMP fold of 12. The results are displayed for a dip of 45° pointing toward the north (i.e., oriented at 0° to each patch) and at a time of 0.4 s. The responses correspond to a number of CMP bins in three parts of the survey:

1) along the central receiver line (Figure 10.15b),
2) along the central source line (Figure 10.15c), and
3) an areal view of the center of the survey (Figure 10.15d).

There are significant amplitude and phase changes indicating that the output of DMO applied to real seismic data collected with this geometry would have a strong geometry imprint for any northerly dipping events of 45° close to a time of 0.4 s. The imprint in this case is mostly due to the excluded shots and receivers, although it is also possible to see a variation in DMO response at each shot and/or receiver line intersection.

Other examples with different geometry and fold indicate similar results. In general, there are problems with steeper dips at shallow times and with dips across narrow patches. These imaging problems point to a need to either:

a) calculate some form of amplitude and phase compensation (inverse DMO operator), or
b) redesign the acquisition geometry to improve the DMO output contributions in all bins for the desired target dip and time. A general rule for areal (land) geometries is to not use narrow patches for DMO applications.

Although these studies specifically involved DMO, it is likely that other imaging techniques (e.g., prestack time or depth migration) suffer from similar problems, namely, insufficient in-phase cross-line contributions that lead to a poorly focused image.

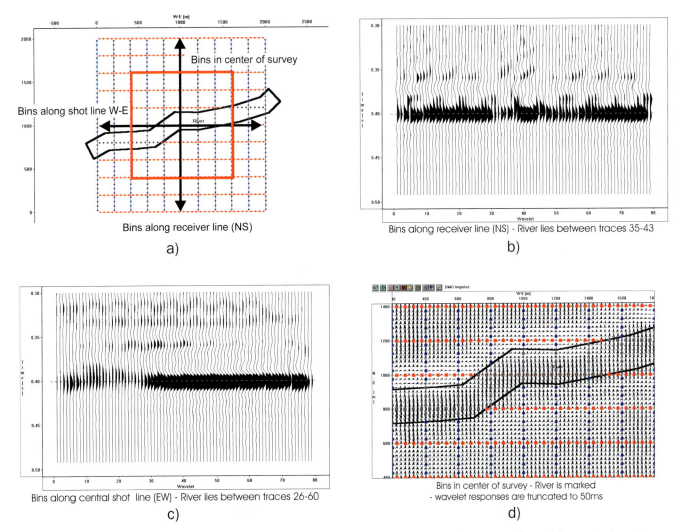

Fig. 10.15. a. 3-D survey showing DMO analysis bins. DMO model response: b. along a portion of the central receiver line, c. along the central source line, d. for all bins in the center of the survey.

10.7 STACK

CMP stacking sums the traces in a CMP bin to increase the S/N. The question, "What is Fold?" is riddled with problems. There can be good fold and bad fold (Figure 10.16). For example, by repeating a source point many times into the same receiver, only ambient (or background) noise is attenuated. This type of fold does not result in improved signal. Reciprocal raypaths, where source points and receivers are interchanged, can also be classified as bad fold. On the other hand, combining traces where energy travels by a variety of different raypaths is better. The combination of such wavefront energies can be considered to be a crude version of Huygens' secondary wavelet summation. This type of stacking can be loosely classified as good fold.

The process of stacking attenuates high frequencies. On top of that, any midpoint smear within the bins is a high-cut filter. Bin scatter can also be detrimental for the imaging of dipping structures.

If a good mix of offsets is present in each CMP bin, not only will random noise be attenuated, but coherent noise such as ground roll and multiple energy will also be attenuated due to the presence of such noise at different times on different offsets. If, however, a good offset mix is lacking, multiples, ground roll, and other coherent noise may be present on the CMP stack traces.

Figure 10.17 shows a synthetic shot with two events at 2400 ms and 3040 ms. There are two linear noise trains with a velocities of 200 m/s and 250 m/s. Using the geometry of CDPs from a real 3-D survey, one can construct synthetic CMP gather data with the traces of each bin determined by the traces of the shot that have the same offset. In other words, the model has flat layers and the noise is the same everywhere.

In the synthetic CMP gathers of Figure 10.18, the linear noise looks somewhat irregular due to the irregular mix of offsets in each CMP bin. When these CMP gathers are subjected to NMO and stacked, the noise is not canceled. Moreover, the noise is attenuated in different ways depending on the offset mix in the various bins. The original linear noise now has a presence on the stack, which may interfere with, or be interpreted as, real data.

10.8 ACQUISITION FOOTPRINTS

Amplitude anomalies that are related to acquisition are often difficult to detect on vertical seismic sections. They are more evident on time slices or amplitude maps of horizon slices. A number of factors are known to enhance the acquisition footprint (La Bella

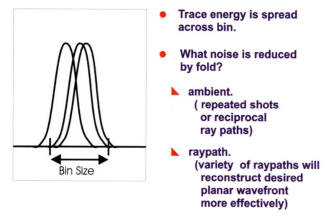

Fig. 10.16. Stacking—good fold versus bad fold.

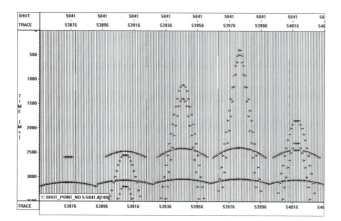

Fig. 10.17. Stack and noise attenuation—raw synthetic 3-D shot with reflection events and linear noise.

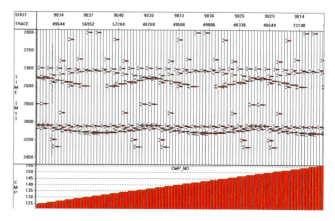

Fig. 10.18 Stack and noise attenuation—raw CMP gathers.

10.9 Migration and Random Sampling

et al., 1998). It is desirable to reduce this footprint as much as possible in the acquisition process. Acquisition footprints can arise from a variety of causes, such as:

- changes in the signal-to-noise ratio caused by bin-to-bin fold variations
- linear noise stacking in different ways because of variation in offset distribution
- multiple energy remaining after stack because of variation in offset and azimuth distribution
- azimuth and/or offset polarization
- DMO failing to create constructive interference due to missing offsets in some bins
- backscatter noise combining in different ways in neighboring bins as a result of azimuth and offset distribution variations

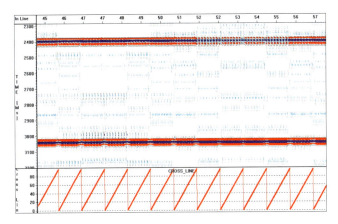

Fig. 10.19a. CMP stack for several in-lines (equivalent of one "Box").

Figure 10.19a shows a CMP stack for several in-lines that are the equivalent of one "box." Note the amplitude variations within each in-line, and the variation from one in-line to the next. Figure 10.19b shows a graph of amplitude of the events at time 2400 ms and time 3040 ms. Note the variations. The deeper event is not as much affected by the noise imprint. Figure 10.19c shows the time slice of the event at time 2400 ms with the true 3-D geometry superimposed. Note how the time slice amplitudes follow the 3-D geometry (source and receiver lines). Such noise remnants after stack are the most common form of "acquisition footprints" (geometry artifacts).

10.9 MIGRATION AND RANDOM SAMPLING

In recent years there have been many papers that investigate how to use random sampling to reduce migration noise caused by coarse spatial sampling. Recommended readings are the papers published by Vermeer (1996) and Schuster and Zhou (1996).

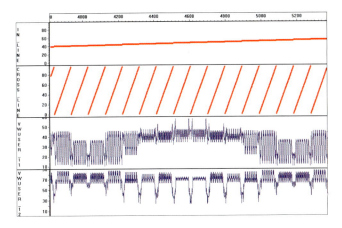

Fig. 10.19b. Graph of amplitude of events at time = 2400 ms and time = 3040 ms.

One of the more interesting illustrations is in Vermeer (1996). Figure 10.20 displays the effect of sampling density for a horizontal event for five wavelets that are formed by stacking along the diffraction curve through the output sample position. The input data are a horizontal event with a wavelet corresponding to a vertical and horizontal resolution of 12.5 m. The migration of the input data sampled every 12.5 m produces a perfect result (left-most wavelet). Increasing the spatial sampling to 33 m produces a wavelet with precursor migration noise (middle wavelet). Random samples with a mean value of

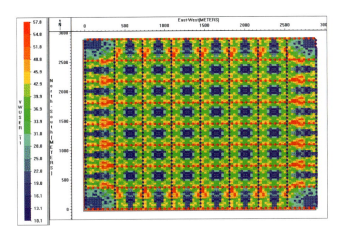

Fig. 10.19c. Time slice of event at t = 2400 ms with geometry superimposed.

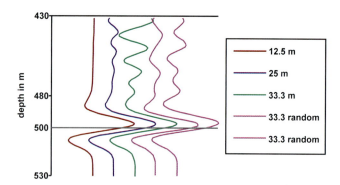

Fig. 10.20. Effect of sampling density on migrated response (after Vermeer, 1996).

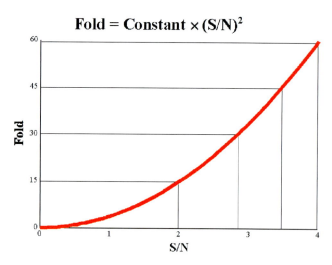

Fig. 10.21. Fold versus signal-to-noise ratio.

33 m and a spread of 11 m either side of 33 m make up the two right-most curves. Clearly the random sampling has reduced the migration noise.

One should try to sample 3-D wavefields as finely as possible to reduce migration noise for a wavelet having the desired resolution. If fine sampling is impossible, then random sampling can be effective. It is possible to achieve somewhere near the desired resolution, and with less migration noise.

For 3-D data, the migration of regularly sampled data already has an element of randomness, because the migration process is a function of the distance from input point to output point. These distances are not regularly distributed in a 3-D geometry. Equalized DMO may be applied to irregularly sampled 3-D data, requiring no particular acquisition geometry other than just source and receiver coordinates (Beasley and Mobley, 1998).

To achieve random sampling in the field, it is not necessary to work at it. Normal bin to bin variations, of standard geometries (orthogonal, nonorthogonal, zig-zag, etc.) provide the necessary randomness. Moved sources and receivers create a further degree of randomness.

10.10 MAKING ADJUSTMENTS FOR DATA QUALITY

If data quality is unexpectedly low, the interpreter, in conjunction with the processor, has to evaluate the data set to decide how to improve the signal-to-noise ratio. A multitude of possibilities exists. The easiest way to improve the S/N is to enlarge the bins. In a traditional 3-D survey this usually means doubling or quadrupling the bin size. The latter choice will increase the fold by a factor of four and double the S/N (Figure 10.21). If many offsets are redundant or the offset mix is poor, the gain in S/N may not be as large as expected. Note that linear source noise is non-Gaussian and therefore the S/N for that particular noise will not improve with increasing fold.

Another option would be to perform some sort of borrowing of information from neighboring bins while retaining the original bin size. This borrowing can be done depending on missing offsets or azimuths in the center bin. These missing offsets can be added from neighboring bins to increase the fold and therefore improve the S/N. The only problem with this approach is the large distance from one central midpoint to the next midpoint where the required information is obtained. However, the data will appear smoother, and in the case of extremely bad data, such a strategy may make a 3-D survey interpretable. It is important to note that this kind of bin borrowing is exactly what DMO performs. Thus, if DMO is included in the seismic processing sequence, there should be no requirement for bin balancing/borrowing.

In extreme cases of bad data, a running mix of information of neighboring bins will smooth the data possibly to the point of making a correct interpretation difficult, because the high frequency content may have been eliminated.

If the data quality is high, one may limit the required offsets and azimuths to get a good representation for each bin. If the fold distribution over the entire 3-D area can be smoothed, then the overall data quality may be as good as or better than the original full-fold survey. In conventional 3-D surveys there is no method of decreasing the bin size to get a better spatial resolution. The Flexi-Bin® approach distributes

10.10 Making Adjustments for Data Quality

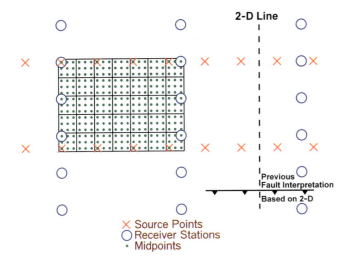

Fig. 10.22a Bins before bin-rotation, with anticipated fault direction.

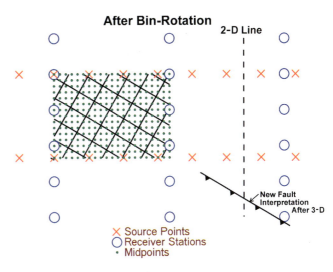

Fig. 10.22b Bins after bin-rotation, with actual fault direction.

the midpoints within each bin in a prescribed manner. This distribution produces sub-bins that can be summed into bins with small bin size increments equal to the sub-bin dimensions rather than in multiples of the natural bin size. If the data quality is marginal, the bin size can be increased; if the data quality is high the bin size can be decreased (Section 5.7).

Bin rotation may be required to combine two or more neighboring 3-Ds or if significant geological features are displaying unexpected trends (Figure 10.22). Such features could include channels, faults, and reef trends. Figure 10.22a shows a 3-D layout following an east-west fault direction based on geology only. The 3-D survey was acquired and revealed that the actual fault direction was northwest-southeast (Figure 10.22b). The bins were rotated to have less bin lines affected by the fault cut. Such bin rotation is difficult to achieve with a conventional 3-D survey because of the central distribution of midpoints within each bin. In a survey where the midpoints are distributed within each bin (e.g., irregular line spacings, crooked line orientation, random, or Flexi-Bin®), such a change in bin orientation may be easier to accomplish.

Chapter 10 Quiz

1. Consider a typical 3-D situation in an area of exploration interest and determine what one would consider to be an ideal set of film displays (if any) for a permanent record.
2. Why is the cross-line dimension of the patch important when considering refraction statics?
3. Give some efficient means of obtaining reflection statics coupling in a 3-D survey.

11

Interpretation

11.1 INTERPRETATION SYSTEMS

Only a few key interpretation issues are addressed in this chapter. Comprehensive discussions on the subject of 3-D seismic interpretation can be found in Brown (1991) and Sheriff (1992). Interpretation is often the last hands-on step in seismic data gathering and analysis. The trend is to bring the interpreter closer to the processor and the acquisition contractor. It is, in fact, important that interpreters be involved in all aspects of 3-D seismic survey design, acquisition, and processing, or at least they should be kept informed of the progress in these areas. Additional data displays or field tests may answer critical questions once there is communication among all people involved in 3-D design through interpretation.

Paper displays should be used as a permanent hard copy record of the 3-D survey (see Chapter 10). Carefully selected paper displays are an inexpensive medium for display and are important for presentations to management and other interested parties. Some things are more easily detected by the human eye when interpreting paper sections than through workstation interpretation. Several software platforms may be required to complete a 3-D seismic interpretation. There are numerous software systems on the market today that can effectively handle 3-D surveys of any size. For small and medium size surveys, PC-based systems may suffice (e.g., Win Pics, Vest, Kingdom 3-D). For larger surveys, and mainly because of the superior speed, one may prefer one of the Unix workstation-based interpretation systems (e.g., Landmark, GeoQuest, or Paradigm). The selection of an interpretation software package is subjective. Each system has unique features that may be good for a particular purpose. Some important features that should be reviewed before purchasing a software system would be: ease of use, contouring options, time-slice viewing, fault handling, integration of well synthetics, 3-D visualization, animation possibilities, the ability to make arbitrary lines through the data volume, import and export options, merging of several 3-D data sets, and processing options such as filtering or phase rotation.

Seismic data can be displayed in many different formats, such as variable intensity (VI), variable area (VA), or a combination of the two as variable intensity and area (VIVA). Variable intensity displays can utilize a number of different color bars to highlight particular features. When horizon times are picked on a workstation, reflection amplitudes can be measured simultaneously or extracted at a later stage. Some software packages allow the user to pick horizon times in both the vertical and horizontal planes. The picks can be displayed for each horizon individually or in combination with others.

11.2 MAPPING

Geological mapping is usually available before any 3-D interpretation begins. These maps provide a useful framework for 3-D-based geophysical mapping. From work with existing 2-D or 3-D data sets, one often knows which horizon, isochron, amplitude, or other attribute maps are needed to define the prospect. The following list is a sample of maps and displays that should be considered (other displays may be available):

time structure
flattened seismic sections
isochron
amplitude
amplitude difference
instantaneous frequency
instantaneous phase

172 Interpretation

time slices
horizon slices
chair displays
3-D visualization
shadowing
transparency displays
coherence cube

Smoothing maps may clarify structural and stratigraphic relationships imaged by a 3-D data volume. Smoothing should be applied with caution to avoid eradicating short-wavelength anomalies. One needs to test different smoothing operators to produce meaningful results. A smoothing window of three traces may work well in some areas, but in others a window of seven (or more) traces may be required. It is also useful to make difference maps. A difference map is the mathematical subtraction of the original map from the smoothed map. Difference maps accentuate subtle features, stratigraphic details, and acquisition artifacts.

Hand-contoured maps may indicate geology more accurately than computer-generated maps. Computer-generated maps should always be made, however, because they reveal detail that may otherwise be ignored. Mapping depth structure remains a significant challenge even with a 3-D data set. Good velocity determinations are essential to correctly image structure. All available well information needs to be carefully integrated. Often a shallow geological marker, if it has a smooth structure, can be used as a datum when this marker can be clearly identified on the seismic image. Combining isochrons to the target horizon with an interval velocity function (areally varying) may produce accurate isopach maps. Reliable depth structure maps can result by adding these isopach maps to shallow datum maps.

11.3 INTEGRATION

Previously acquired 2-D and 3-D data should be fully integrated with the current 3-D data and with geologic and engineering control. Integration with any available geological mapping is essential for the best prospect interpretation. At times, seismic data mapping alone does not provide the information necessary to commence drilling. However, when fully integrated with geologic, engineering, and production control, one can substantially increase the value of a prospect. Even old fields experience significant production increases with this integrated approach and follow-up drilling.

Data integration includes different disciplines. Geologic input is most important and includes regional structure mapping and reservoir-scale interpretation. Hydrodynamic mapping can have a significant impact on migration paths for hydrocarbons and helps define where compartment boundaries may exist in the 3-D image. In purely exploratory areas, questions about source rocks and seals should be addressed before a major 3-D survey is contemplated. Engineering data can provide valuable definitions of high-production and low-production seismic facies and help site new drilling locations. Additional enhancement of the data, e.g., 3-D inversion, time-to-depth conversion, AVO analysis, etc., can sometimes provide information that may further reduce drilling risk significantly.

Acquisition, processing, and interpretation of several 3-D surveys in one area require careful planning of the data merge to minimize differences between the surveys. With fewer differences, the interpretation and further drilling activities have a greater chance of success. One such data merging has been undertaken in Oman and is described in Ligtendag (1999). When faced with an ever-increasing 3-D coverage in that country, a large-scale 3-D merge was undertaken. It consisted of three major steps: (1) include prestack traces of previous 3-D surveys in the processing of new surveys; (2) include a rim of stacked data of previous 3-D surveys in the migration of new surveys; and (3) merge the new and old surveys on the interpretation workstation. This megaproject included approximately 30 000 km^2 of migrated seismic data. Amplitude and phase corrections, redatuming, regridding, and rotation were undertaken where appropriate. In addition, a megagrid was laid out to unify the location identification for all 3-D surveys. Future surveys are to be harmonized into the same system. This project in Oman has shown that such data integration into one large reference framework is possible and feasible, even though the acquisition parameters were not consistent throughout the surveys.

Regional high-resolution aeromagnetic (HRAM) surveys can supplement several 3-D surveys by linking areas of data coverage over areas where no 3-D data were recorded. The advent of the airborne magnetometer increased the use of magnetics in exploration for oil, gas, and minerals from almost nonexistence at the end of World War II to its current status as a key geophysical method for the discovery and delineation of new sedimentary basins and additional deep sections in known basins throughout the globe (Steenland, 1998). GPS navigation, lower altitude drape flying, and modern analysis techniques

(Rhodes and Peirce, 1999) allow not only deeper basement structures to be identified, but interpretable magnetic sources in the sedimentary section are being imaged (Biegert and Millegan, 1998). The observed magnetization along fault planes detected by HRAM data is explained by vertical fluid flow, which involves the transport of iron in oxidized waters followed by precipitation of exotic iron-bearing minerals during the ascent of the water as it undergoes redox reactions (Peirce et al., 1998).

11.4 ACQUISITION FOOTPRINTS

The interpreter should always look for acquisition footprints in the data when analyzing 3-D images. If there is a clear relationship between the acquisition geometry and trends or anomalies in the interpreted maps, one should attempt to remove these nongeological artifacts in processing, if possible.

The acquisition geometry should always be available as an overlay on any interpretation map. Such correlation is easiest to notice on amplitude maps and time slices. Since acquisition footprints are worse in the shallower section and at lower fold, it may be necessary to review fold maps at equivalent time slices in order to research any correlation between them. Once a problem has been encountered, the minimum and maximum offset, unique offset and unique fold, as well as the migration apron displays, may help clarify the problem.

11.5 SEISMIC ATTRIBUTES

A good summary and classification of seismic attributes has been published (Chen and Sidney, 1997) that details event-based and volume-based attributes. Examples for an event-based attribute might be instantaneous frequency or reflection strength, while a volume-based attribute might be the coherence cube as a measure of structural discontinuities. Numerous software packages allow the simultaneous display of several mapped attributes in overlay or even in a blended display.

Exxon has attempted to model the feasibility of utilizing attributes from different 3-D surveys over the same area (Lewis, 1997). Lewis points out that repeat surveys must be registered with the old surveys in three dimensions to reduce artifacts. However the alternative of subtracting attribute maps that measure amplitude or frequency in a time interval of the surveys removes that necessity (see Chapter 12). The differences of attributes can simplify the analysis of monitoring surveys. Forward modeling is suggested to select the best attributes that can best characterize the reservoir units and any displaced fluids.

11.6 GEOSTATISTICS

Geostatistics provides an excellent method for integrating a sparsely sampled well data set with a densely sampled seismic data set. Variogram analyses can estimate the spatial correlation and the anisotropy of the data, which helps to interpolate data. Reservoir parameters, available from logs and/or core samples, are considered as the error-free primary data set. Kriging methods estimate unknown values of the primary data set between wells, while honoring the known values at the existing well locations (Trappe and Hellmich, 1998). The more densely sampled secondary data (seismic data and their attributes) are used to guide the interpolation in the cokriging process. The output still honors the well information while taking advantage of the attributes of the denser seismic grid. Since several solutions are possible, it is often practical to simulate many different cokriging results and calculating an average. Parameter and error ranges are estimated with more confidence. Interpretation problems may arise if one of the data sets is undersampled, either spatially or simply with an insufficient number of samples.

Geostatistics is a proven tool for integrating well and seismic information through the spatial analysis of reservoir parameters and seismic attributes. The responsibility of the geoscientist is to ensure that the results truly represent the data and are consistent with the geologic model (Hirsche et al., 1998).

11.7 IMMERSION TECHNOLOGY

Since the subject of this book is 3-D design, it seems only appropriate to discuss the advent of the truly three-dimensional workspace. The most widely accepted terms are the "immersive environment" or "immersion technology." The most dramatic effect has been the spatial sensation that one feels through projection of the seismic data onto three-dimensional surfaces. Numerous solutions are offered at largely varying prices. Large 3-D surveys may very well warrant such expenditure with the increased level of collaborative interpretation to be reached.

The display surfaces may be large (on the scale of a room) and either flat or curved. Domes (or half-domes) may utilize only one projector (front lit), while large curved screens may require several projectors. Square boxes that one can walk into (as if in a cave) are back lit, and the data are projected onto the sides, floor, and roof.

Distortions do occur with most of these systems to some degree. Selection of the most suitable system will largely depend on the number of interpreters that will work with the data and system, personal preference, and of course price.

12

Special Interest Topics

12.1 DIGITAL ORTHOMAPS AND GIS DATA

Digital orthomaps (DOMS) are a new tool in the planning phase of a 3-D seismic survey (Crow, 1994). These maps are made by digitizing high-quality aerial photographs of a project area and then sending a global positioning satellite (GPS) crew to the prospect to record the GPS coordinates of several features that can be visually identified on the photos. The digital images are then rectified to agree with the GPS shots taken across the image space. These rectified images can then be transferred to 3-D design software packages in digital form, such as a TIFF format. The software tools for the planning of 3-D seismic grids allow integration of this digital imagery in the planning of the seismic program. Skids and offsets can be positioned on the image to avoid obstacles seen in the digital image (Figure 12.1).

Aerial photography and even video can be integrated with GPS information and historical survey information to provide extremely accurate and consistent survey information. GPS technology can provide XY accuracy for selected reference points within 1 cm ($\frac{1}{2}$ in), while the Z coordinate is about a factor of two less accurate. Historical data such as well and pipeline locations and cultural features should be integrated into 3-D design as much as possible. The accuracy of 3-D base maps increases significantly when these steps are taken. In the context of a 3-D survey, such reduction in location uncertainties is important and necessary.

DOM files are generally supplied in TIFF format (or geo-referenced TIFF), but may also come in GIF or JPEG format. Such files are raster data. A vector file can also be used.

In recent years many 3-D planners have incorporated GIS (Geographical Information Systems) data files into the planning process. Such "SHAPE" files contain vector data (polygons) and associated database information for each polygon. Many planners also use other vector files (e.g., Autocad files—DXF). The uses of GIS can be:

permit tracking and information
line clearing information
tracking damages
tracking health and safety requirements
drilling statistics
production statistics
vibrator tracking
general GIS queries (e.g., How many source points lie on slopes of more than 15°?)

12.2 TRANSITION ZONES

Transition zones are generally defined as shallow marine, lake, marsh, and river environments. Transition zone surveys usually have problems that are more related to land acquisition than marine acquisition. Mixed sources (e.g., vibrator and dynamite) are commonly used in environmentally sensitive transition zone areas. In contrast, true marine surveys are conducted on the open ocean where large areas can be covered with ships that tow large source arrays and several long streamers.

The fewer variables that one has to introduce into transition zone acquisition, the better. One type of source for the entire survey (e.g., dynamite) is preferable to a variety of sources (e.g., vibroseis, dynamite, mud gun, air gun, and water gun). Similarly, using one type of receiver for the entire survey is far more preferable (but, again, often impossible) to using land geophones, marsh phones, and hydrophones in respective areas of the survey. Often a variety of re-

176 *Special Interest Topics*

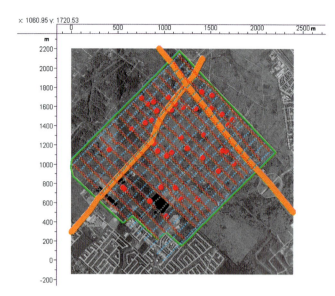

Fig. 12.1. Three-dimensional design over a digital image.

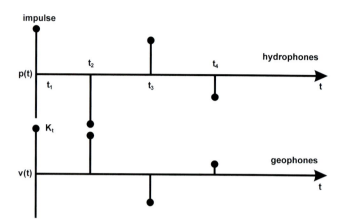

Fig. 12.2. Hydrophone versus geophone response. Impulses at times t_2, t_3, and t_4 are water reverberations (after Barr and Sanders, 1989).

ceivers are required to record the appropriate data. Because of the variables involved, testing is more intricate in transition zone surveys than in true land surveys. A dynamite shot will look different on monitors of land geophones than on those of hydrophones. An airgun shot also has a different appearance on different receivers.

Phase matching needs to be considered when processing data from different areas of a transition zone survey. Theoretical phase curves are an important guide for the processor. In practice, data may look different than the theoretical prediction. Phase differences may actually change as the water depth increases. Winter versus summer acquisition needs to be considered. Winter may provide more efficient crew movement on the ice, but data quality may not be as good as that which is recorded in non-ice environments. Ice may introduce flexure noise that makes signal not discernible. If the water is ice covered, then the sources and the receivers should be placed below the ice surface. Summer acquisition may interfere with the tourism industry, but data production may be higher and data quality better. Environmental constraints may limit the options for equipment usage and timing of the seismic program.

Dual-sensor data acquisition with ocean bottom cables (OBC) reduces ghosting effects on transition data (Barr and Sanders, 1989), thereby enhancing the frequency content of the data. By plotting water-column reverberations over time for hydrophones and geophones (Figure 12.2), one can see that summing after appropriate scaling cancels the ghosts. Many process-

ing centers have proprietary software for such purposes. OBC data can now be acquired in water depths up to 2000 m.

The use of a universal recording system (such as the Fairfield BOX) simplifies transition zone acquisition. In shallow water (1–30 m), a combined geophone/hydrophone package can be used at each station similar to that used in OBC; in muddy or swamp areas, marsh phones may be used; and on dry land, regular geophones may be planted. Thus with a universal recording system, a single shot can be recorded at all these different types of receiver stations. It is desirable to use a real-time radio mode of acquisition to simplify the layout logistics.

12.3 PRESTACK TIME AND DEPTH MIGRATION

In complex-structure environments, it may be necessary to perform a prestack time migration on the data set. An accurate velocity model is not as essential as it is for poststack migration. In prestack time migration the data are stacked after the migration has been applied in an attempt to obtain a better positioning of the reflector in space. Several iterations may be required; unfortunately, two to three iterations of prestack migration are expensive. Hopefully, this expense is outweighed by the additional accuracy that this process provides.

Figure 12.3 shows schematically the difference in stacking between poststack and prestack migration. In poststack migration (Figure 12.3a) the midpoints that are assumed to be scattered about the bin are stacked first for central midpoints. The midpoints that fall within the migration aperture can then be migrated. In prestack migration (Figure 12.3b), all the midpoints that fall within the migration aperture are migrated before being stacked. The energy contributions of poststack versus prestack migration can be quite different. Prestack time migration is getting to be a standard deliverable for many processing centers.

Prestack depth migration is a successful solution to the 3-D seismic imaging problem in situations where there is a severe lateral variation in velocity across the 3-D image space. Experience with this technique indicates that the quality of the depth image is sensitive to the accuracy of the velocity model (Canning and Gardner, 1993).

The economic benefits of prestack time or depth migration need to be considered (Schultz, 1997). Are the imaging improvements so significant that they warrant the additional expense? One might consider a staged approach to prestack imaging by comparing partial data volumes after prestack and poststack migration.

12.4 TIME-LAPSE (4-D) SEISMIC

With time-lapsed (4-D) data sets, one opens numerous possibilities for data enhancement. Not only can one evaluate the movement of reservoir fluids, but time-lapsed data sets can be combined to give new geologic insights. Hesthammer and Løkkebø (1997) have demonstrated on a data set from the North Sea that studying the difference cube may be a good measure of the relative amplitude of the signal, while multiplying dip maps may eliminate noisy areas of the surveys and highlight faults. Using the larger absolute amplitudes of the two surveys provided the best data set for detailed seismic interpretation because the continuity of real reflections was enhanced and coherent noise was reduced. The formula "If $|A| > |B|$, then A, else B" offered a particularly sharp image. Even a straight summing of the amplitudes appeared to be better than any single data set. In this example from the Gullfaks Field, the processing parameters were chosen to be the same for each 3-D data set; different processing streams may very well reduce the improvements that were realized. Many problems may be overcome by combining the data prestack and then processing the super-gathers.

An excellent treatment of the topic of time-lapse data and its use in reservoir management has been provided by Jack (1997). A technical risk spreadsheet provides an objective evaluation of the merits of 4-D projects (Lumley et al., 1997).

12.5 CONVERTED WAVE 3-D DESIGN

Three-component 3-D surveys are useful for detecting fractured reservoirs or any other type of geology where one can expect anisotropic behavior of rock

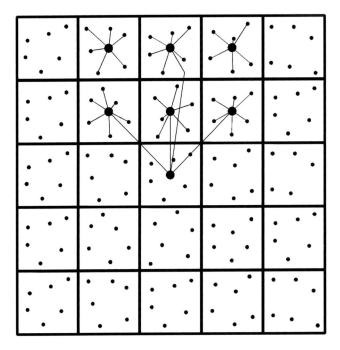

Fig. 12.3a. Poststack migration.

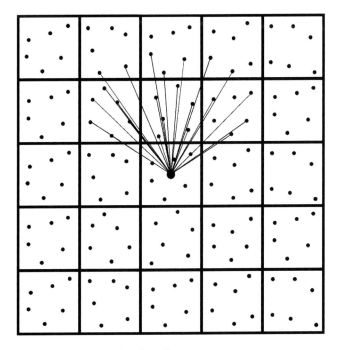

Fig. 12.3b. Prestack migration.

properties. When designing converted wave 3-D surveys, one has to calculate a suitable range of offsets where the converted waves will be present. The common conversion point (CCP), where the incident P-wave is converted to the reflected S-wave, does not lie at the midpoint position between source and receiver. Rather this conversion point is offset from the source position a distance X_c (Figure 12.4) given by

$$X_c = \frac{r}{1 + \frac{V_S}{V_P}}, \quad (12.1)$$

where r is the distance between source and receiver, V_S is the average S-wave velocity, and V_P is the average P-wave velocity. Hence the illumination area depends upon the V_P/V_S ratio.

The main question for converted wave 3-D design is how to bin the data that do not have their reflection points at the CMP positions. Several possible solutions have been introduced. Fold calculations and binning during data processing may be based on a bin size defined by Lawton (1993):

$$B = \frac{RI}{1 + \frac{V_S}{V_P}}, \quad (12.2)$$

where RI is the receiver interval.

Example: assume $\dfrac{V_P}{V_S} = 2$

then $X_c = \dfrac{r}{1.5} = \dfrac{2}{3} \times r,$

and $B = \dfrac{RI}{1.5} = \dfrac{2}{3} \times RI.$

The above formulas are based on the asymptotic conversion point concept (Figure 12.4) rather than on depth-variant CCP mapping that would be somewhat more complicated. Although V_P/V_S has to be known to determine the bin size, one does not have to know the velocity ratio prior to acquisition of the data. The results from the initial processing determine the ratio. V_P/V_S is >1 and usually has a value of ~2.0. Therefore a converted-wave bin size is always larger than a standard 3-D common-midpoint bin (a V_P/V_S value of 1 results in the normal bin size of half the receiver station interval RI). If the CCP bin size is not changed to reflect this conversion point location, then the CCP fold distribution map has a striped appearance, possibly with holes in it (Figure 12.5a). This striping effect can be reduced by selecting the source line intervals to be an odd integer of the receiver interval rather than an even integer. In addition, when increasing the bin size as described above, the fold map will be even smoother (Figure 12.5b).

Ideally, one would like to find a field geometry that produces a smooth midpoint distribution in the compressional data and also a smooth CCP distribution in the converted-wave stacks while using a constant bin size for both. Cordsen and Lawton (1996) developed such a technique in the planning for the Blackfoot 3-C 3-D survey acquired by the CREWES Project in 1995. In the following discussion of this design concept, the V_P/V_S ratio is assumed to be constant at 2.0. When

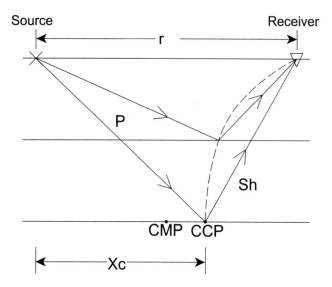

Fig. 12.4. Converted-wave raypaths. P = P-wave, Sh = shear wave.

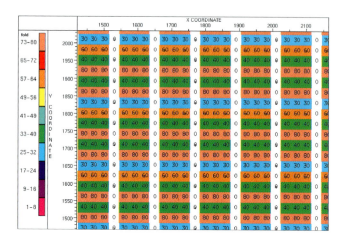

Fig. 12.5a. Fold distribution for converted S-wave component with regular bins.

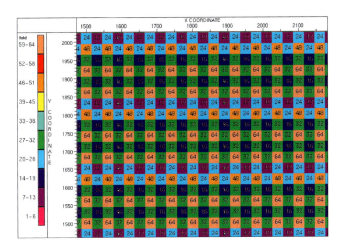

Fig. 12.5b. Fold distribution for converted *S*-wave component with large bins.

significant variations in this ratio occur, the CCP distribution should be checked with a commercially available software package. Note that the calculation of the bin size and the CCP distribution in the receiver-line direction are different from that in the source-line direction.

The CCP distribution in the receiver-line direction can be determined by choosing a source-line interval according to the following formula:

$$SLI = RI \times i. \qquad (12.3)$$

If i = even integer, the CCPs occur at intervals of

$$\frac{RI}{1 + \frac{V_S}{V_P}};$$

if i = odd integer, the CCPs occur at intervals of

$$\frac{RI}{2\left(1 + \frac{V_S}{V_P}\right)};$$

and if i = (integer + 0.5), the CCPs occur at intervals of

$$\frac{RI}{4\left(1 + \frac{V_S}{V_P}\right)}.$$

The CCP distribution in the source-line direction can be determined by choosing a receiver-line interval according to the following formula:

$$RLI = SI \times j. \qquad (12.4)$$

If j = any integer, the CCPs occur at intervals of

$$\frac{SI}{2\left(1 + \frac{V_S}{V_P}\right)};$$

if j = (integer + 0.5), the CCPs occur at intervals of

$$\frac{SI}{2\left(1 + \frac{V_S}{V_P}\right)};$$

and if j = (integer ± 0.25), the CCPs occur at intervals of

$$\frac{SI}{4\left(1 + \frac{V_S}{V_P}\right)}.$$

Figure 12.6a shows the even (and identical) fold distribution accomplished for both the compressional and converted-wave stacks using the relationships i = (integer + 0.5) and j = (integer ± 0.25). The acquisition parameters were as follows (Cordsen and Lawton, 1996):

Receiver interval	60 m
Source interval	60 m
Receiver line interval	195 m
Source line interval	270 m
Patch	12 × 54 channels

Figure 12.6b indicates the CCP distribution for the converted-wave stack when using a V_P/V_S ratio of 2.0. For the converted wave stack the CCPs fall onto a 10 × 10 m grid. For the compressional-wave stack, the midpoint distribution is on a 7.5 × 15 m grid (Figure 12.6c). Both of these grids allow summing at the larger 30 × 30 m bin size as indicated.

Sometimes it is hard to understand the variations in the V_P/V_S ratio prior to data acquisition. Under such circumstances one may want to randomize the station locations further by including one or more of the following methods: smaller line or station spacings, Flexi-Bin (as in the above example) and/or staggered lines (Figure 12.7), nonorthogonal or totally random station positions (see Section 5.14). The resulting randomization in the midpoint distribution creates a smooth fold distribution that is almost independent of the V_P/V_S ratio.

Vermeer (1999b) points out the asymmetric illumination of the converted-wave acquisition and concludes that parallel geometries are far better than orthogonal geometries. On the other hand, parallel geometries may not be suitable for the detection and analysis of azimuth-dependent effects.

180 Special Interest Topics

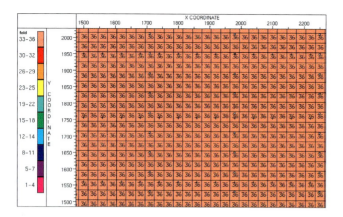

Fig. 12.6a. *P*-wave and converted *S*-wave fold.

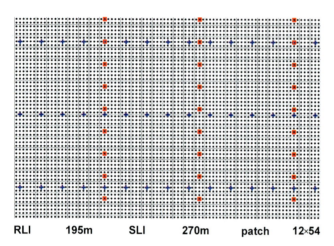

RLI   195m   SLI   270m   patch   12×54

Fig. 12.6b. Converted *S*-wave CCPs.

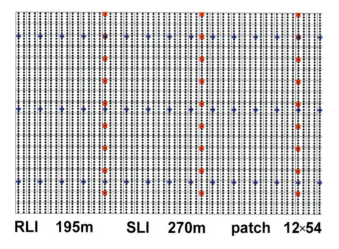

RLI   195m   SLI   270m   patch   12×54

Fig. 12.6c. *P*-wave CMPs.

Processing flows of converted-wave 3-D data are still in a development phase, and one should have sufficient knowledge of the design and processing procedures to take advantage of the additional infor-

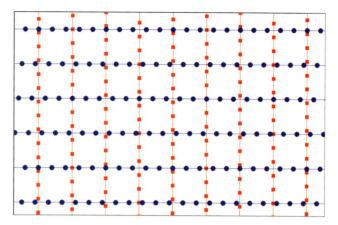

Fig. 12.7. Randomization of station and line locations for converted-wave acquisition.

mation gained by recording three data components. Processing of 2-D converted wave data is discussed in detail by Hauser (1991). Converted-wave data exhibit amplitude changes at far offsets. These amplitude variations can be interpreted as AVO effects only if changes in reflection coefficients are properly considered as the raypaths approach the critical angle.

12.6 3-D INVERSIONS

By using inversion technology, one can expect to obtain a better understanding of the different effects that thickness and impedance changes of stratigraphic units have on seismic reflection character. Three-D inversion can also provide the additional perspective of porosity distribution as long as the lithology is known. A 3-D impedance cube, in addition to a conventional seismic volume, can benefit 3-D interpretation and reservoir characterization (Duboz et al., 1998). Thin beds can be defined that are difficult to detect in the input 3-D seismic volume (Figure 12.8). This knowledge is extremely helpful when planning horizontal well bore trajectories.

Some important data-processing issues that must be considered for inversion calculations are the following:

1) True reflection amplitudes must be maintained in the processing sequence.
2) A short-offset stack version of the data set should be made to avoid AVO effects.

Well ties are extremely important in 3-D inversion. One starts by estimating wavelets near the well ties over a few selected traces. Experience has shown that approximately 20% of the well ties may yield odd wavelets. It is important to consider the average fit of

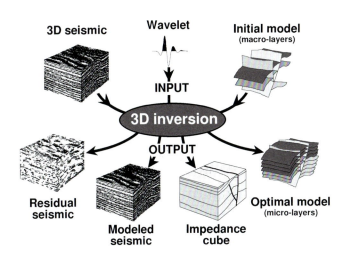

Fig. 12.8. Data input and output for 3-D inversion (Duboz et al., 1998).

all well ties before making a final phase determination and adjustment of the 3-D data. If there is a good explanation for the phase changes of the wavelets observed over an area, an interpolation technique should be used to create a smooth phase transition across the total 3-D image space.

Since impedances are estimated in the inversion, wells that have both sonic and density logs are best suited for calibrating the inversion results. In a 3-D survey in which a variety of sources were used, one must perform source phase matching before attempting inversion. Failing to do this will result in an incorrect inversion, because phase ties may be more erratic than in a survey where only one source was employed. Ideally the same well, or similar logs, should be used to tie the different sources in a survey for the determination of the phase correction operator.

12.7 FUTURE DIRECTIONS

New field hardware will open possibilities for new layout strategies. In particular some of the positioning strategies (GPS) and cableless systems (telemetry) will mean considerably more flexibility in laying out receiver positions. Light-weight, solid-state, three-component geophones will allow the deployment of a much larger number of geophones than is presently possible.

Through new software to help analyze processing requirements, one will be able to experiment with strategies that use some of these new receiver layouts, and therefore possibly different source layouts. Such analyses can determine the factors that drive the survey design to optimum resolution limits and also save time and money.

A new method of reservoir analysis deals with tuning cubes (Partyka et al., 1999). These are 3-D data cubes that are limited in spatial and temporal coordinates to the zone and area of interest (limited by horizon picks). The time-domain 3-D data are transformed into the frequency domain using a discrete Fourier transform. This spectral decomposition allows the data to be viewed as frequency slices, highlighting thin beds and geological discontinuities.

A technique called voxel tracking has become available with the advent of volume rendering and visualization. A voxel is a "volume element," i.e., in a 3-D seismic data volume it is a data sample. Voxel trackers follow a three-dimensional path from a seed sample through the data volume (Dorn, 1998).

The future directions that 3-D seismic technology may take can be described in terms of the task that needs to be done and the tools that will be available to accomplish that task.

The Task: To meet or exceed geophysical requirements (resolution) at lower cost.

The Tools:
  New acquisition technologies

- New layout strategies
- Light-weight multicomponent geophones (solid state)
- Telemetry
- GPS
- 3-C recording
- 9-C recording
- Time lapse (4-D) recording

  New processing methods

- Improved velocity analysis
- Prestack time migration
- Prestack depth migration
- 3-C processing
- Time lapse (4-D) processing

  New interpretation methods

- 3-D AVO and AVA
- Geostatistics
- Reservoir monitoring
- Attribute analysis
- Coherence cubes
- Tuning cubes
- Voxel tracking
- Sensory feedback systems:
  Vision (immersion technology)
  Touch (haptic feedback)
  Audio (data sonification)

Answers to Quiz Questions

Chapter 1 Quiz Answers

1. The receiver line interval (*RLI*) is the distance between adjacent receiver lines.
2. Migration apron is the width of the fringe area that needs to be added to the 3-D survey to allow proper migration of any dipping event or diffraction point at the edge of the image area. Migration aprons may be different in strike and dip directions. Note that the migration apron is not the same as the fold taper needed to build up fold at the edge of the survey. The fold taper and migration apron usually overlap.
3. Calculate the diagonal of the box formed by adjacent receiver and source lines. This number reflects the minimum offset for the CMP bin exactly in the center of that box. It is the largest minimum offset of the 3-D survey.
4. A super bin is an area large enough to incorporate a sufficient number of traces to have a good representation of offsets for velocity determination. It should not be so large that the geology varies within its boundaries. There is no one particular size that can be predetermined, but a common choice is 3 × 3 bins.

Chapter 2 Quiz Answers

1. Which of the following factors affect in-line fold and cross-line fold, assuming no other changes in patch geometry?

			In-line	Cross-line
a.	X_{max}	Maximum offset	(X)	(X)
b.	*RLI*	Receiver line interval	( )	( ) within usable X_{mute}
c.	*NRL*	Number of receiver lines	( )	(X)
d.	*SLI*	Source line interval	(X)	( )
e.	B_s, B_r	Bin size	(X)	(X)
f.	*NC*	Number of channels	(X)	(X) depends on layout

Chapter 3 Quiz Answers

1. A narrow azimuth 3-D survey is better for AVO work because of the longer offsets and their linear distribution.
2. The migration apron should be at least equal to the width of the Fresnel zone radius (containing 70% of the diffracted energy) and up to $Z \times \tan 30° = 0.58 \times Z$ to capture 95% of the diffracted energy.
3. Not necessarily, but often the migration apron may be the same on all sides of the survey. The migration apron depends on the larger of: (1) the lateral movement of dipping surfaces from migration and (2) the width necessary to capture most of the diffraction tails. The allowance for lateral movement is $MA = Z \times \tan(\theta)$, where θ is the apparent dip in the direction being considered. For diffractions, use a 30° scattering angle, $MA = Z \times \tan(30°) = 0.58 \times Z$. If the dips in one direction exceed 30° (and one wishes to resolve those dips), then the migration apron in the dip direction may be wider than in the strike direction.

    Another consideration is how wide a fold taper is needed to build up fold. Depending on the design, the fold taper width may be different in the in-line and cross-line directions. A judgment call based on costs and the detailed interpretive objectives must be made to determine how much of this fold taper overlaps the migration apron.
4. The image area has full fold, fully migrated data. The migration apron area has only partially migrated data and a stack with full fold. The fold taper zone is the area necessary to build up fold in the stack data. Often, there is some overlap of the migration apron and the fold taper.

Chapter 5 Quiz Answers

1. In an orthogonal design, the bin in the center of a box has the largest minimum offset X_{min}.

183

2. X_{min} in the brick pattern depends on the source and receiver line spacings; however, one can increase the receiver line interval of the orthogonal design without sacrificing X_{min} as much as in the orthogonal design. The X_{min} distribution should be confirmed with a 3-D design package.
3. The basic concept of Flexi-Bin® is that the ratios of *RLI/SI* and *SLI/RI* are noninteger values. The line intervals are arranged in such a manner that the midpoints are evenly distributed within a bin, rather than all falling onto one central midpoint.
4. One must have excellent access for the application of the zig-zag method (e.g., in the desert).
5. In both survey types the receiver lines radiate out from a center. In a star survey the source points are along the receiver lines. In a radial survey the source points are along concentric circles.

Chapter 6 Quiz Answers

1. In dynamite acquisition the effort is determined by charge size, charge depth, and number of holes per shot location.
2. In Vibroseis acquisition, the expression pad time refers to the total length of time that the vibrator pad rests on the ground while sweeping, i.e.,

   pad time = (sweep length) × (number of sweeps)

3. See figure 6.27

Chapter 7 Quiz Answers

1. When the water depth is shallow, one may want to consider marsh phones that can still be pole planted. When the water depth exceeds about 2 m *(5 ft)*, then hydrophones are a better choice. These may be strung out on the bottom by dropping the cables from a boat, or they may be planted by divers.
2. The operator in the recording truck checks the cable connections electronically. The line crews should fix poor connections or leakages. Any shutdown of the system requires additional stabilizing and should be avoided.
3. Distributed systems still have some cable connections between the boxes and the central recording unit, while telemetry systems do not.
4. BOX or RSR systems allow the storage of many records in the box near the geophone groups. The stored information may then be downloaded to the main recording truck once or twice per week with the help of a data collection unit (DCU).

Chapter 9 Quiz Answers

1. A final plan shows the source and receiver lines in their final positions (as actually used at the time of recording) as well as any access and detour routes. A preplot shows the anticipated source and receiver line prior to surveying in the field.
2. A script file must contain all information necessary to describe the source and receiver locations and their relative geometry for all source points of a 3-D survey.
3. Roll-on and roll-off lowers fold. For the useful offsets though, there may not be any difference whatsoever. The higher fold that results when not using roll-on/off may be useful only if all far offsets are included in the final stack. If this is the case, the 3-D survey is probably not properly designed, because the patch is then not large enough in the middle of the survey.
4. Snaking is used when only one shooting crew is available. Snaking over several lines is more time efficient if extra equipment is available to do that option. Ping-ponging is used when two or more shooting crews are available. They can travel parallel on several source lines or in-line along the same source line.

Chapter 10 Quiz Answers

1. Permanent film displays depend on each company's objective and may vary from play to play. Full-scale displays for the entire section (to basement) every tenth (or so) line may be useful. In addition, displays of every line or every second line for the zone of interest, may prove to be good for reviewing and picking well locations, especially in the absence of workstation access.
2. The cross-line dimension needs to be sufficiently large to pick up measurements of shallow refractors. Also, one needs to ensure that refractions from deeper horizons fall within the cross-line dimension.
2. Reflection statics coupling is achieved by:
   a) breaking up the regularity of receiver or source line spacing
   b) moving sources and/or receivers
   c) adding source lines in the receiver line direction
   d) adding receiver lines in the source line direction

Glossary of terms related to 3-D survey design

3-C 3-D A 3-D survey which is acquired using a standard source and 3-C geophones.

3-component geophone (3C) A geophone with three orthogonal sensors. The phone must be planted with known orientation, usually one component in-line, one cross-line, and one vertical.

3-D symmetric sampling Symmetric sampling applied in 3-D surveys.

4-C receiver station A receiver station with a 3-C geophone plus a hydrophone.

4-D survey See **time-lapse survey.**

9-C 3-D A 3-D survey which is acquired using three sources: a standard source, an in-line shear source, and a cross-line shear source. Each wavefield generated by each source is then recorded with 3-C geophones.

achievable resolution A lower resolution than potential resolution caused by noise (multiples, ground-roll, and ambient noise) and by irregular or coarse sampling.

acquisition imprint Imprint of 3-D source and receiver geometry onto 3-D data and data attributes.

active receiver (station) A receiver (station) belonging to the group of receivers (stations) that are recording data.

actual resolution A lower resolution than achievable resolution caused by various suboptimal processing steps (errors in velocity model, phase errors, etc.).

air blast The pressure wave that travels through the air from the source to the geophone.

air gun A marine energy source that creates seismic wavefields by releasing compressed air.

air-gun array A collection of air guns optimized to generate a sharp source wavelet.

alias-free sampling Sampling that introduces no improper frequency or wavelength information into 3-D data.

ambient noise Noise produced by the environment (engines, people, wind) in contrast to source generated noise.

amplitude striping A geometry imprint typical for streamer surveys, may be caused by feathering but also by width of multisource multistreamer geometry.

amplitude versus azimuth (AVA) Variation in reflection amplitude as a function of source-to-receiver azimuth.

amplitude variations with offset (AVO) Variations of reflection amplitude as a function of offset distance. Behavior depends on Poisson's ratio of rocks at the reflecting interface.

anisotropy Variations in rock properties as a function of direction of analysis.

antiparallel acquisition Sailing adjacent boatpasses in opposite directions.

aperture A range of illumination angles used in migration.

apron The width of the fringe area that needs to be added to a 3-D survey to allow proper migration of any dipping event. Although this is a distance rather than an angle, it has been commonly referred to as the (migration) aperture. Other synonyms are **migration apron** and **halo.**

areal geometry Acquisition geometry consisting of a dense (sparse) areal grid of receiver stations and a sparse (dense) areal grid of sources.

array (pattern) A geometrical arrangement of sources and/or receivers used to suppress noise of certain wavelengths.

array length Number of elements times the distance between the elements.

array response The amplitude response of an array as a function of wavelength and direction.

aspect ratio The ratio of the narrow dimension of a rectangle divided by the wide dimension. In 3-D design, the ratio of the cross-line dimension divided by the in-line dimension.

azimuth Angular direction in degrees relative to north.

basic sampling interval The sampling interval required for alias-free sampling of the whole continuous wavefield (including ground roll).

basic signal sampling interval The sampling interval required for alias-free sampling of the desired part of the continuous wavefield.

basic subset The 3-D subset of an acquisition geometry consisting of traces that have smoothly varying spatial attributes. Possible subsets could be cross-spread, common-offset gather, CMP gather, etc.

Beylkin's formula The formula for the computation of spatial resolution (Beylkin, 1985).

bin An area used to gather traces with midpoints that fall inside that area. Bins can be any shape but usually are square or rectangular.

bin fractionation An implementation of the orthogonal geometry that intentionally creates subgroups of midpoints within a natural bin. Uses shifts in station positions between adjacent acquisition lines to create the effect.

bin interval The distance between adjacent bins.

bin rotation Reprocessing 3-D data to create in-line and cross-line orientations that are different than those involved in the original data. Normally used with bin fractionation or when combining (interpreting) two or more 3-D surveys where processing used different bin grid angles.

bin size The area of a bin. Normally determined by source and receiver station spacings.

binate Literally, take every second sample. Often used as reducing the number of traces by taking every n^{th} sample, n not necessarily being 2. Because this is a resampling operation leading to larger sampling intervals, a spatial alias-filter should be applied before binary. See also decimate.

bird-dog A slang term referring to the client's quality-control person that oversees an acquisition crew.

boat pass A single crossing of the survey area by a seismic vessel in multisource, multistreamer acquisition.

box The area of a 3-D survey bounded by two adjacent receiver lines and two adjacent source lines. A box contains all the statistics that describe the middle of the 3-D survey. See **unit cell.**

brick or brick-wall geometry An orthogonal geometry in which the source lines are staggered between receiver lines. Used to reduce the X_{min} of a geometry for better shallow coverage.

button A tightly grouped arrangement of receivers in a button patch geometry.

button patch geometry A 3-D geometry patented by Arco in which the receivers are laid out in buttons and sources are positioned around the buttons. Rather than a dense areal grid as in the areal geometry, this geometry has a checkerboard pattern of dense receivers and empty spaces.

cable The wire connecting receiver groups to the line units.

cat push A slang term referring to the person who supervises line cutting. Derives from Caterpillar bulldozers.

center-spread acquisition Acquisition with as many receivers to the right of the source station as to the left. May also apply to receiver stations where each receiver station has an equal number of sources on either side. Center-spread acquisition creates symmetry in common-source gathers and in common-receiver gathers.

chair display An interpretive display in which the 3-D volume is sliced into two depth sections and a time section connected in a chair shape.

charge The amount of dynamite (lb or kg) used for one source point, sometimes consisting of several **shotholes.**

circular design A design that uses **circular patches.**

circular patch A patch with an outer edge that approximates a circle.

CMP fold The theoretical fold calculated by binning CMPs.

COCA gather Common-offset/common-azimuth gather.

common conversion point (CCP) In converted-wave (*PS*) acquisition, the CCP is the equivalent of the **common midpoint.** It is the point between the source and receiver where the downgoing *P*-wave generates an upgoing *S*-wave.

common depth point (CDP) The common reflection point for dipping reflectors and complex velocity fields.

common midpoint (CMP) The theoretical reflection point that lies midway between a source and receiver. Assumes no structural dip and no unusual velocity variations exist. A group of traces that shares the same midpoint.

common scatter point (CSP) A way of analyzing prestack data (Bancroft and Geiger, 1994). A reflecting surface is thought of as a specular surface, with each point generating a set of diffractions. The processing collapses these diffractions prestack to gather the energy to the appropriate point in the subsurface.

common-offset gather One of the basic subsets of parallel geometry. For a perfect subset, the source/receiver azimuth should be constant in the gather, but usually it is not.

common-offset stack A display in which traces with the same offset from different sources have been stacked. The traces are displayed after the application of NMO. This display is useful for determining mute, detecting multiples, and initial analysis of AVO.

continuous fold Coverage in the case of proper sampling; reflection points can be reconstructed for any position within a properly sampled data set.

conversion point The point in the subsurface where conversion of *P*-waves to *S*-waves occurs. Normally this point is about two-thirds the offset, as compared with 1/2 for *P-P* reflections.

converted wave An *S*-wave generated when a portion of a *P*-wave is converted into shear-wave energy at a reflecting surface.

convolution A mathematical process applied to two or more time series that corresponds to multiplication of the Fourier transforms of the time series.

correlation The vibrator sends a programmed wavetrain, or chirp, into the subsurface. Each reflection is also a chirp. In the correlation step, the reflected trace is correlated with the chirp to convert each chirp reflection to a short, compact wavelet.

critical reflection The angle at which waves are refracted instead of reflected.

cross-line In a 3-D survey this is the direction orthogonal to the receiver lines. It is usually the same as the direction of the source lines.

cross-line offset The component of offset that is in the cross-line direction, perpendicular to the receiver lines.

cross-line roll A patch (swath) move perpendicular to the receiver lines.

cross-spread One of the basic subsets of orthogonal geometry. Consists of all traces that have a source line and a receiver line in common.

decimate Literally: take every tenth sample. Often refers to reducing the number of traces by taking every n^{th} sample, n not necessarily being 10. As this is a resampling operation leading to larger sampling intervals, a spatial alias-filter should be applied before decimation. See also **binate.**

deconvolution ("decon") A mathematical process that collapses wavelet signatures into sharp spikes. Commonly used in processing to boost high frequencies. Inverse of convolution.

density The mass per unit volume of rock. Usually measured in kg/m^2 or g/cm^3. Density has some effect on seismic velocity.

depth migration Seismic migration performed in the depth domain rather than the time domain.

depth structure map A map of a particular horizon where the vertical dimension is depth.

diffraction A seismic event generated by a scattering point.

diffraction traveltime surface The collection of traveltimes associated with a diffraction event.

dip moveout (DMO) The change in reflection time due to change in position of reflection point for CMP traces. The processing step that attempts to move reflections to a common-reflection point. When combined with NMO correction, the construction of zero-offset traces from offset traces.

dip shooting The use of a geometry oriented in the main dip direction.

dip/strike decision The decision whether to choose dip or strike direction as the shooting direction in marine surveys and in swath surveys.

direct wave A wave travelling in the surface layer directly from source to receiver.

distributed system A 3-D data acquisition system. The signals from several receiver groups are collected at a remote line unit and then transmitted to the recording truck (cf. **telemetry system**).

doghouse A slang term for the recording truck.

double zig-zag geometry A 3-D geometry involving two zig-zag paths for the source lines.

drag The amount of movement of vibrators between each shake within the source array at one source station (cf., **move-up**).

dual-sensor technique OBC with a geophone (array) and a hydrophone (array) in every receiver station.

dwell In nonlinear vibrator sweeps, the dwell is the additional sweep effort applied at higher frequencies; usually quoted as dB/octave.

edge management Optimization of image area, migration apron, fold taper zone, DMO, and cost considerations to arrive at an efficient design at survey edges.

effective spread length The product of number of stations and station interval. Should be used in all 3-D design computations rather than spread length. A similar definition applies to linear arrays with equidistant elements.

effort A general term for the amount of vibrator energy put into the ground. Determined by the number of vibrators, peak ground force, sweep length, and the number of sweeps.

exclusion area An area that is not accessible because of natural or manmade hazards or a no-permit area.

far offset The farthest offset recorded. Used to refer to farthest offset in a particular patch.

Farr diagrams Named after Gordon Farr (Gulf Canada). Small windows (100 ms or so around the first-breaks) of shots are displayed en echelon in a surface consistent position. All traces have LMO applied. Thus all traces recorded at a common receiver will line up vertically. Geometry errors are obvious with such displays.

feathering The deviation of towed streamer from track followed by vessel and source.

feathering angle The average angle between source track and streamer direction.

final survey plan The plan after the survey is recorded, with all skids and offsets entered. This plat is often what must be submitted to the regulatory authorities.

Flexi-Bin geometry An implementation of orthogonal geometry that creates subgroups of midpoints within a natural bin. Uses line spacings that are nonintegers of station spacings to create the effect.

fold or fold-of-coverage Usually the number of traces in a bin. Sometimes the number of overlapping basic subsets of a geometry.

fold rate The increase in fold (either in-line or cross-line) per line interval.

fold taper zone The area around a 3-D survey in which the fold increases from zero to full-fold.

fold-of-illumination Usually the number of raypaths hitting a bin on a reflector. Better—the number of overlapping illumination areas of basic subsets.

footprint of geometry a. The imprint of acquisition geometry on 3-D data displays. b. The illumination pattern on reflector that resembles geometry of source and receiver lines.

fracture porosity Reservoir porosity created by cracks or fractures in the rock, sometimes enlarged by subsequent dissolution. Commonly sought as a potential gas reservoir, particularly in carbonates. Oriented fractures will polarize shear waves into fast (parallel to fractures) and slow (transverse to fractures) components.

frequency The number of oscillations per second, expressed in Hertz. Units = s^{-1}.

Fresnel zone The first Fresnel zone is the area around a reflection point within which constructive interference occurs. Often used instead of **zone of influence.**

full-swath roll An implementation technique for large surveys in which the whole swath is moved by the full width of the swath when the cross-line roll is done.

full-fold area The area of the survey where full fold is achieved, neglecting the effects of DMO or migration (cf., **image area**).

gather A collection of seismic traces.

geometry imprint May occur in two disturbing ways: periodic, reflecting the periodicity in the acquisition geometry, and nonperiodic, such as striping in marine acquisition caused by feathering and multisource, multistreamer acquisition.

geophone A sensor that records the particle velocity created by seismic waves.

geophone array The geophones laid out at a receiver station to achieve a desired array response.

geophone group Each receiver station is usually occupied by several geophones in a group to improve signal-to-noise ratio. The geophones in a group are laid out to form a geophone array.

geophysical trespass In some areas it is illegal to record a geophysical measurement of any kind over another owner's mineral rights without a permit from that owner.

geostatistics A mathematical technique of crosscorrelating areally distributed data sets. Can be used for time-to-depth conversion by correlating well-control and seismic data.

global positioning system (GPS) A satellite positioning system based on calculating the range to at least four satellites. Most accurate mode of operation is "differential GPS" which can give x,y accuracy of 1–2 m, and z accuracy of 5–10 m at 1 s rate. Greater accuracy can be achieved by repeating observations. Tree cover or rough topography can obscure the signal.

gravity coupling Coupling of geophones to the earth using gravity only (most OBS techniques use gravity coupling).

ground force The amount of force exerted by a vibrator (cf., **peak force**).

ground roll The surface wave generated by a source. These are high-amplitude, low-velocity waves. Often, the fastest non-P-wave is also called ground roll whereas it is often a first-arrival shear wave.

group interval See **receiver interval**.

halo A term sometimes used to mean fold taper zone.

horizon A particular reflecting surface or its reflection.

horizon slice An interpretive display in which the displayed surface follows an interpreted horizon.

hydrophone An underwater sensor that measures pressure changes instead of the particle velocity measured by a geophone.

illumination area The area on a reflector covered by all traces in a basic subset.

image area The portion of a 3-D survey that has full-fold data after DMO and migration.

impedance The product of bulk density and wave velocity. In equation form,
impedance Z = density × velocity.
Reflection coefficient = $(Z_1 - Z_2) \div (Z_1 + Z_2)$.

in-line The direction parallel to the receiver lines in a 3-D survey.

in-line offset The offset in the direction parallel to the receiver lines.

in-line roll The movement of a swath in the direction parallel to the receiver lines. Typically, inline rolls are only a few stations and are accomplished electronically.

interleaved acquisition A technique using overlapping boat passes to compensate for large streamer distances. The distances are large to avoid streamer entanglement.

inversion Seismic inversion is a mathematical process that calculates the impedance contrasts that produce the observed seismic response. The process and the results are nonunique.

isotropic A condition in which a rock system has the same rock properties in all directions.

largest minimum offset (LMOS) The largest X_{min} of all the bins in a box or in some statistically complete subset of the 3-D survey. The maximum shortest offset (see X_{min}).

lateral resolution The minimum distance over which two separate reflecting points may be distinguished. Primarily a function of frequency.

linear moveout (LMO) A static shift applied to each trace equal to offset/velocity, where "offset" is the source-to-receiver distance and "velocity" is normally chosen to be the approximate first-break velocity. The effect of the shift is to move all first-breaks close to zero time (i.e., flatten on the first-breaks).

line geometry Acquisition geometry in which sources and receivers are arranged along straight acquisition lines.

line turn The change of direction of the seismic vessel in preparation for the next boat pass.

live receiver An active receiver ready to record data.

LMOS Largest minimum offset.

marsh phone A geophone designed to be used in marshy conditions. It must be planted in the marsh bottom and can be immersed.

maxibin or macrobin The neighborhood of bins used for velocity analysis or for time picking in statics determination.

maximum cross-line offset The maximum offset in the cross-line direction.

maximum in-line offset The maximum offset in the in-line direction.

maximum recorded offset The largest offset recorded in a swath.

maximum unaliased frequency The highest frequency that can be recorded in a basic subset without creating aliased frequencies because the trace spacing is too large.

MESA A 3-D survey design package available from Green Mountain Geophysics.

midpoint A point halfway between a source station and a receiver station.

midpoint scatter The common situation in 3-D acquisition where the midpoints of traces that contributes to a bin are spread out across the bin, rather than concentrated in the center of the bin.

midpoint/offset coordinate system A coordinate system based on midpoint coordinates and offset vectors. A simple transformation allows conversion to the source/receiver coordinate system.

migration A process in seismic processing in which reflections are moved to their correct reflection points in space.

migration aperture The range of illumination angles used in migration.

migration apron The additional distance that must be added to each side of a 3-D survey to ensure that the migration process can work.

migration noise Noise created by the process of migration due to irregular or coarse sampling.

migration stretch Vertical distortion of a seismic wavelet caused by the movement of reflection energy to a potential reflection point. The distortion is a function of offset (in prestack migration) and dip.

minimal data set A single-fold 3-D data set that is suitable for migration (a basic subset, excluding the 2-D line).

minimum resolvable distance The smallest distance between two events that can be resolved.

move-up The distance that vibrators must move between the last sweep of one source point and the first sweep of the next source point.

multicomponent recording An acquisition technique that records two or more components of the seismic wavefield.

multiple Seismic energy that has been reflected more than once.

multiple suppression Any process that reduces preferentially the energy of multiple arrivals.

multisource, multistreamer acquisition The use of one or more source arrays in combination with many (4 to 12) streamers.

mute function In a common-source gather, energy beyond certain offsets is discarded because it becomes distorted by refractions and other effects. The offsets that are retained increase with depth. The mute function is the increase of usable offsets as a function of two-way traveltime.

narrow azimuth geometry A 3-D geometry that has a small aspect ratio. This geometry means most of the recorded energy comes from a narrow cone of azimuths oriented parallel to the long axis of the survey.

natural bin A bin with dimensions of (1/2 source station spacing) × (1/2 receiver station spacing).

near-offset trace A trace recorded with a relatively short source-receiver distance.

near-trace cube A 3-D data set extracted from a 3-D survey using only the nearest offsets. Used for quality control.

NMO discrimination Using the amount of normal moveout observed to characterize events by their velocity.

NMO stretch Vertical distortion of a seismic wavelet caused by NMO correction. The distortion is a function of offset.

no-permit area An area of a 3-D survey which is excluded because a permit could not be obtained for surface access.

nominal fold Full fold using all receivers, all offsets, and natural bins.

nonorthogonal geometry Any 3-D geometry that does not use a rectilinear grid of lines. Used to refer to line geometries in which the source lines are not orthogonal to the receiver lines.

normal moveout (NMO) The variation in reflection time as a function of source-receiver distance (offset).

OBC Ocean-bottom cable.

OBS Ocean-bottom seismometer.

ocean-bottom cable technique (OBC) Marine acquisition technique using receiver cables laid out on the sea-floor. Receivers may be inside or outside the cable. Usually have dual sensors, but 4-C OBCs are available as well.

ocean-bottom seismometer (OBS) Self-contained receiving and recording unit. Used mostly by academia but also being tried for exploration.

offset a. The distance between a source group center and a receiver group center for a particular trace. b. Sometimes used to refer to stations that are moved a short distance perpendicular to the line, usually because of access difficulties.

offset distribution May mean two different things: distribution of offsets in a CMP or bin (also called offset sampling), or the distribution of offsets across the bins. Preferably, both distributions should cover the whole range of offsets occurring in the geometry, whereas the offset intervals should be irregular. Regularity in the CMPs may lead to aliasing of multiples for low-fold data; regularity across the bins may lead to visible periodic geometry imprint.

offset sampling The sampling of offsets within a CMP or bin.

OMNI A 3-D survey design package available from Seismic Image Software Ltd.

orthogonal geometry Acquisition geometry with parallel source lines running perpendicular to parallel receiver lines.

pad time The sweep length times the number of sweeps for a vibrator source design.

parallel geometry Acquisition geometry with parallel source lines running parallel to parallel receiver lines.

patch In an orthogonal survey, a rectangle of receivers that are spread over several receiver lines. Several sources may have the same patch. The patch moves around the survey for different source points.

patch shooting A method of OBC data acquisition. Several receiver lines are deployed and stationary; source lines are acquired with source points inside and outside the patch. All lines are picked up in one roll and moved to the next location.

peak force The maximum amount of force that a particular vibrator is designed to apply to the ground.

peg-leg multiple Multiples caused by horizons that are relatively close together. The short time delay of the peg-leg event makes the velocity of the peg-leg multiple close to the velocity of the primary events, and therefore harder to separate and suppress.

phase The argument of a wave function. If $y = \sin(\omega t)$, ωt is the phase, expressed in degrees or radians.

Poisson's ratio The ratio of transverse strain to longitudinal strain, usually denoted by σ. It is one of the elastic constants that affects both P- and S-wave velocity.

porosity Pore volume per unit volume, expressed as a percentage.

potential resolution Theoretically best possible resolution. Can be computed using Beylkin's formula.

prestack depth migration A migration process applied in the depth domain (instead of the traveltime domain) to unstacked traces. This process uses ray tracing to compute the diffraction traveltimes and can cope with complex velocity models.

prestack process Any process applied before all traces from a particular CMP or bin are summed together.

prestack time migration A migration process applied in the time domain (instead of depth) to unstacked traces. This process uses the double square-root equation to compute the diffraction traveltimes.

prism waves Raypaths with a double bounce: against reflector and against the flank of a salt dome before returning to the surface.

proper sampling A data set is properly sampled if the underlying continuous wavefield can be faithfully reconstructed from the sampled values.

P-wave This is the type of elastic body wave normally considered in seismic work. The particle motion is in the direction of wave propagation.

quasi-random sampling A method of irregular sampling to reduce migration noise caused by coarse sampling.

radial Target-oriented acquisition geometry used for known salt domes; source points are along concentric circles.

random geometry Geometry with a random distribution of source and receiver locations.

ray-trace modeling Modeling that computes raypaths as they pass through each layer.

receiver The recording device in a seismic survey.

receiver interval The distance between each group of receivers.

receiver line The line along which receivers are laid out in a straight-line 3-D survey. Receiver lines are parallel to the in-line direction.

receiver line interval The distance between receiver lines measured orthogonal to the receiver lines.

receiver station A group of geophones linked by a wire.

reciprocity theorem The assumption that interchanging the position of the source and the receiver will lead to the same recorded trace.

regular sampling Sampling with a constant sampling rate.

resolution The ability to discriminate between closely spaced subsurface features.

ROV Remotely operated vehicle. Used for subsea operations.

running mix A summing of traces in which the number of traces summed is larger than the number of traces advanced between each calculation.

salvo The number of source points taken before the patch must be moved, i.e., the number of source points in a template.

script file The computer file that tells the recording system the geometry of each template in the survey.

SEG-P1 format An SEG-approved standard format for recording positioning data.

SEG-Y format An SEG-approved format for recording seismic data. There are many variations of the SEG-Y format, so it is often necessary to test for compatibility between different systems.

seismic reservoir monitoring Monitoring the production of hydrocarbons using seismic techniques (repeat surveys or time-lapse surveys).

semblance A measure of multichannel coherence, usually measured as a function of stacking velocity. The correct stacking velocity should produce the most coherence and the highest semblance.

SH wave The horizontal component of motion in a shear wave.

shear wave A body wave in which the wave motion is transverse to the direction of propagation.

shot A dynamite charge used as a source in a seismic survey. Shot is often used to refer to any seismic energy source (see source point).

shothole The hole drilled to contain an explosive charge. Shallow holes should be below the weathering layer and deep enough not to blow out. Deep holes reduce ground roll.

signal-to-noise ratio (S/N) The power of the desired energy (signal) divided by the remaining energy (noise).

similarity tests Checking to make sure that all the vibrators in an array are in-phase.

skids Sometimes used to refer to stations moved a short distance along the line, usually because of access difficulties.

slowness The inverse of velocity.

sonic log A well log of seismic traveltime. The frequencies used in a sonic log are much higher than those in seismic data.

source The point of energy release in a 3-D survey. The usual sources are dynamite or vibrators on land and airguns in water. This term is preferred over **shot** in this book.

source density The number of sources per unit area, usually expressed as sources per km^2.

source interval The distance between adjacent sources in a 3-D survey.

source line (shotline) The line along which source points or vibrator points are placed, usually at regular intervals.

source line interval The distance between source lines, usually measured perpendicular to the source lines.

source point A location of a source (shot).

source/receiver coordinate system A coordinate system based on source and receiver coordinates. A simple transformation allows conversion to the midpoint/offset coordinate system.

source-generated noise That part of the seismic wavefield that needs to be removed in processing.

source-receiver pair The receiver array and the source point that produce a given recorded trace.

spatial continuity The absence of spatial irregularities, such as edges, missing source points, missing receivers. Slow variation of spatial attributes of all traces in a data set.

spatial frequency The wavenumber.

spider diagram A diagram used to display azimuth distribution in a 3-D design package. Each leg of the spider points in the direction from the source to the receiver, and the length of the leg is proportional to the offset.

SPOT imagery A French satellite imaging system that produces black and white images with the highest resolution available currently.

spread An arrangement of receivers associated with a source point. In a cross-spread, the line of receivers forms the receiver spread and the line of source points forms the source spread.

spread length The distance between ends of a spread.

SPS format A standard SEG format proposed by Shell for writing script files that contain comprehensive information about the geometry of the survey.

stack array The combination of geophone arrays and regular equidistant offset sampling in a CMP. In 3-D, the stack array concept can be applied only in parallel or full-fold 3-D geometry

stack response Response as a function of wavenumber computed for all offsets that contribute to the stack in a CMP or bin.

stack section A time section produced by stacking without application of migration.

STAR Target-oriented acquisition geometry used for known salt domes. Source points are along receiver lines that are distributed as spokes of a wheel.

static coupling The static correction for each receiver is based on many source paths into that receiver. If a direct path can be drawn from any receiver to a midpoint, and from there to all other receivers (via more midpoints), then the static corrections are said to be coupled and produce a single solution. In a standard orthogonal geometry there are usually several sets of connected receivers that are not linked to each other, which lead to several independent statics solutions unless macro-bins are used.

statics The time corrections applied to compensate for the slow velocities and elevation differences of the surface weathering layer(s).

stationary-receiver system A marine acquisition system with receivers in fixed position during data recording.

straight-line geometry Any 3-D geometry that uses straight lines for receivers and sources. Source lines are often, but not necessarily, orthogonal to receiver lines.

streamer acquisition A marine acquisition technique using towed seismic cables.

strike shooting The use of a geometry oriented in the main strike direction.

sub-bin In bin-fractionation techniques, a smaller group of traces than the natural bin.

SUMIC A subsea seismic 4-C technique developed by Statoil.

super bin The neighborhood of bins used in velocity analysis.

surface area The area enclosed by the outermost sources and receivers in a 3-D survey.

SV wave The vertical component of motion in a shear wave.

swath a. Width of the area over which the sources are being shot without any cross-line rolls, often with many in-line rolls in one swath. At the end of a swath there is a cross-line roll to set up the next swath (see also patch). b. The collection of all receiver lines laid out at one time. c. A single boat-pass, or a group of adjacent boat-passes, all acquired in the same direction.

swath survey In a swath survey, source lines are parallel to the receiver lines (parallel geometry). Since parallel receiver lines record simultaneously from one parallel source line, swath lines are created midway between source and receiver lines (terminology in use for land and OBC surveys trying to mimic marine multistreamer surveys).

S-wave Shear wave.

sweep The input from a vibrator. Frequencies are varied ("swept") in a precise manner over several seconds.

sweep length The time needed to sweep across the entire frequency band of the sweep.

sweep rate The frequency band of the sweep divided by sweep time. Units are Hz/s (or, more properly, s^{-2}).

symmetric sampling A seismic sampling technique that applies the same sampling for sources and receivers, because the properties of the seismic wavefield are the same in common-source gathers and in common-receiver gathers.

takeout The electrical connection in a receiver cable where a group of receivers is attached.

target depth The depth of the prospective horizon for which the 3-D survey is being designed.

target size The lateral dimensions of the prospective geological reservoir. In 3-D design, the smallest of these dimensions needs to be resolved.

target-oriented acquisition geometry Acquisition geometry optimized for a known geologic structure, e.g., a concentric circle shoot around salt domes.

telemetry system A 3-D recording system that uses a radio system to relay the recorded information from the receiver groups to the recording truck.

template The collection of active receiver stations plus the associated source points.

TIFF file A particular computer format (Tagged Image File Format) commonly used for scanned images.

time slice A map of a seismic attribute at the same two-way traveltime.

time structure map A map of a particular reflector in two-way traveltime.

time-depth function For a given point (particularly for a well), a set of two-way traveltimes and their equivalent depths (true vertical depths), or the mathematical function which approximates such a set of time-depth pairs.

time-lapse survey The repeated acquisition of the same survey area as a tool in seismic reservoir monitoring; also called a 4-D survey.

total nominal fold or full fold The fold calculated for a 3-D survey assuming that all possible offsets are recorded and used.

transition zone An area around a water-land boundary in which neither land nor marine acquisition techniques may be used without special adaptations. Examples include surf zone, large marshes, small lakes, mangrove swamps.

umbilical The pressure hose linking the compressor on a vessel to an airgun array.

uncorrelated record A recorded trace from a vibrator survey in which the input waveform of the vibrator has not yet been removed from the data.

undershooting Most common in marine streamer acquisition where two boats are used to obtain coverage below an obstacle. On land, examples of undershooting include imaging under rivers, towns, etc.

unit cell The area defined by two adjacent source lines and two adjacent receiver lines in an orthogonal geometry. (See **box**.)

vari-sweep A technique for enhancing specific frequency bands by sweeping over narrow frequency ranges and summing later.

velocity control point A point in a seismic survey where velocity analysis has been done.

velocity distribution A list of (time, velocity) pairs for a given location.

velocity model The description of subsurface properties in terms of velocities and velocity boundaries.

vertical hydrophone cable The arrangement of approximately 12 hydrophones strung along a vertical cable attached to the sea-floor and kept vertical by a buoy.

vertical resolution The minimum vertical separation that can be resolved in a seismic survey, expressed either in terms of traveltime or distance.

VHC Vertical hydrophone cable.

vibe A slang term for a vibrator seismic source.

vibrator A seismic source in which the weight of a specially designed heavy vehicle is supported by a central pad and then hydraulically shaken in a precisely prescribed set of varying frequencies. Often several vibrators are used together.

vibroseis A seismic method in which a vibrator is used as the energy source.

VSP Vertical seismic profile. A seismic survey which combines a surface source and downhole receivers. Better name: **WSP**.

walkaway VSP A VSP with downhole receivers and sources at various offsets from the well.

warp Wide-angle refraction and reflection profiling.

wavelength The distance between two similar points on successive waveforms or on a wave train of a single frequency.

wavelet A seismic pulse.

wavenumber The number of wavelengths per unit of distance.

weathering layer A zone of low-velocity along the surface.

well tie The correlation between the seismic interpretation of a particular horizon and the occurrence of that same horizon in a well as interpreted from well logs.

wide-angle profiling (warp) A technique using a very large range of source offsets from the OBS receivers. A basin-analysis tool.

wide-azimuth geometry A 3-D survey geometry that has a broad range of azimuths recorded by most of the receivers. Large aspect ratio (close to square) patches give wide azimuth ranges.

wood gator A large truck-mounted wood chipper used in South Texas to clear brush on seismic lines.

WSP Well seismic profile. A better alternative name for **VSP** because wells are not always vertical and **walkaway VSP** is a contradiction in terms.

X_{max} The continuous maximum offset recorded in a particular 3-D design.

X_{min} The largest minimum offset recorded for most templates in a particular 3-D design. The magnitude of X_{min} directly influences how well shallow reflectors can be imaged. (See **LMOS**.)

X_{mute} The mute distance for a particular reflector. Any traces beyond this distance do not contribute to the stack at the reflector depth. X_{mute} varies with two-way traveltime.

zero offset When a receiver and source are coincident, there is no horizontal distance between them and they are said to have zero offset.

zig-zag geometry A 3-D geometry in which the source points follow a zig-zag pattern between each pair of adjacent receiver lines.

zipper design A 3-D layout strategy for large surveys which uses overlapping swaths.

zone of influence The area around a reflection point within which interference occurs. The size of this area depends on the length of the source wavelet. Not to be confused with the (first) **Fresnel zone**.

zone of interest The range of traveltimes or depths that encompasses the prospective horizons.

REFERENCES

Ak, M. A., Eiken, O., and Fatti, J. L., 1987, Discussion of the stack-array concept continues: The Leading Edge, **6**, no. 8, 28–32. (See also Anstey, N., 1987, Reply: The Leading Edge, **6**, no. 8, 32, 48.)

Allen, K. P., Johnson, M. L., and May, J. S., 1998, High fidelity vibratory seismic (HFVS) method for acquiring seismic data: 68th Ann. Internat. Mtg., Soc. Expl. Geophys., Expanded Abstracts, 140–146.

Anstey, N., 1986a, Whatever happened to ground roll?: The Leading Edge, **5**, no. 3, 40–45. (with letters in Signals section: The Leading Edge, **5**, no. 6, 6, 10.)

——1986b, Field techniques for high resolution: The Leading Edge, **5**, no. 4, 26–34.

——1986c, A reply by Nigel Anstey: The Leading Edge, **5**, no. 12, 19–21.

——1987, Reply: The Leading Edge, **6**, no. 7, 32, 48. (See also Ak, M. A., Eiken, O., and Fatti, J. L., 1987, Discussion of the stack-array concept continues: The Leading Edge, **6**, no. 7, 28–32.)

——1989, Stack-array discussion continues: The Leading Edge, **8**, no. 3, 24–31.

Aylor, W. K., 1995, Business performance and value of exploitation 3-D seismic: The Leading Edge, **14**, 797–801.

—— 1997, The role of 3-D seismic in a world-class turnaround: 67th Ann. Internat. Mtg., Soc. Expl. Geophys., Expanded Abstracts, 725–729.

Bancroft, J. C., and Geiger, H. D., 1994, Equivalent offsets and CRP gathers for prestack migration: 64th Ann. Internat. Mtg., Soc. Expl. Geophys., Expanded Abstracts, 672–675.

Bardan, V., 1997, A hexagonal sampling grid for 3D recording and processing of 3D seismic data: Geophys. Prosp., **45**, 819–830.

Barr, F. J., and Sanders, J. I., 1989, Attenuation of water-column reverberations using pressure and velocity detectors in a water-bottom cable: 59th Ann. Internat. Mtg., Soc. Expl. Geophys., Expanded Abstracts, 653–656.

Beasley, C. J., 1993, Quality assurance of spatial sampling for DMO: 63rd Ann. Internat. Mtg., Soc. Expl. Geophys., Expanded Abstracts, 544–547.

Beasley, C. J., and Klotz, R., 1992, Equalization of DMO for irregular spatial sampling: 62nd Ann. Internat. Mtg., Soc. Expl. Geophys., Expanded Abstracts, 970–973.

Beasley, C. J., and Mobley, E., 1998, Spatial aliasing of 3-D DMO: The Leading Edge, **17**, 1590–1594.

Bee, M. F., Bearden, J. M., Herkenhoff, E. F., Supiyanto, H., and Koestoer, B., 1994, Efficient 3D seismic surveys in a jungle environment: First Break, **12**, no. 5, 253–259.

Beylkin, G., 1985, Imaging of discontinuities in the inverse scattering problem by inversion of a causal generalized Radon transform: J. Math. Phys., **26**, 99–108.

Biegert, E. K., and Millegan, P. S., 1998, Beyond recon: The new world of gravity and magnetics: The Leading Edge, **17**, 41–42.

Bouska, J., 1995, Cut to the quick: investigating the effects of reduced surface sampling in 3D data acquisition: Ann. Mtg. of the Can. Soc. Expl. Geoph., Expanded Abstracts, 181–182.

——1998a, The other side of fold: The Leading Edge, **17**, 31–35.

——1998b, Double vision for interpreters: Case histories showing the value of dual processing for 3-D surveys: The Leading Edge, **17**, 1520–1540.

Bremner, D. L., Crews, G. A., and Musser, J. A., 1990, Method for conducting three-dimensional subsurface seismic surveys: United States Patent 4 930 110.

Brown, A. R., 1991, Interpretation of three-dimensional seismic data, third edition: AAPG Memoir 42, Am. Assn. Petr. Geol.

Brühl, M., Vermeer, G. J. O., and Kiehn, M., 1996, Fresnel zones for broadband data: Geophysics, **61**, 600–604.

Cain, G., Cambois, G., Géhin, M., and Hall, R., 1998, Reducing risk in seismic acquisition and interpretation of complex targets using a Gocad-based 3D modeling tool: Presented at 68th Ann. Mtg., Soc. Expl. Geophys.

Canning, A., and Gardner, G. H. F., 1993, Two-pass 3-D pre-stack depth migration: 63rd Ann. Internat. Mtg., Soc. Expl. Geophys., Expanded Abstracts, 892–894.

Chen, Q., and Sidney, S., 1997, Seismic attribute technology for reservoir forecasting and monitoring: The Leading Edge, **16**, 445–450.

Claerbout, J. F., 1985, Imaging the earth's interior: Blackwell Publications, **18**.

Connelly D. L., and Galbraith, J. M., 1995, 3-D design with DMO modelling: Ann. Mtg., Can. Soc. Expl. Geoph., Expanded Abstracts, 98–99.

Cordsen, A., 1993a, Flexi-Bin™ 3D seismic acquisition method: Presented at the Ann. Mtg. of the Can. Soc. Expl. Geoph.

——1993b, Flexi-Bin™ 3D seismic acquisition in Southern Alberta: Presented at the Ann. Mtg. of the Can. Soc. Expl. Geoph.

——1995a, Arrangement of source and receiver lines for three-dimensional seismic data acquisition: United States Patent 5 402 391.

——1995b, How to find the optimum 3D fold: Ann. Mtg., Can. Soc. Expl. Geoph., Expanded Abstracts, 96–97.

——1999, How much randomness is good for 3-D design?: 69th Ann. Mtg., Soc. Expl. Geophys., Expanded Abstracts.

Cordsen, A., and Lawton, D.C., 1996, Designing 3-component 3D seismic surveys: 66th Ann. Internat. Mtg., Soc. Expl. Geophys., Expanded Abstracts, 81–83.

Crouzy, E., and Pion, J., 1993a, Total petroleum land 3D seismic survey simulation: Presented at the SEG Summer Workshop.

—— 1993b, 3-D DMO simulation in land survey design using 2-D seismic data: 63rd Ann. Internat. Mtg., Soc. Expl. Geophys., Expanded Abstracts, 548–551.

Crow, B., 1994, Integrating acquisition of digital orthomaps (DOMS) into the E & P planning cycle: Presented at the MESA Technology Conference.

Denham, L. R., 1980, What is horizontal resolution?: Presented at the Ann. Mtg., Can. Soc. Expl. Geophys.

Deregowski, S. M., 1982, Dip moveout and reflection point dispersal: Geophys. Prosp. **30**, no. 3, 318–322.

Donze, T. W., Crews, J., 2000, Moving shots on a 3-D seismic survey: the good, the bad, and the ugly (or how to shoot seismic without shooting yourself in the foot!): The Leading Edge, **19**, No. 5, 480–483.

Dorn, G. A., 1998, Modern 3-D seismic interpretation: The Leading Edge, **17**, 1262–1272.

Duboz, P., Lafet, Y., and Mougenot, D., 1998, Moving to a layered impedance cube: Advantages of 3D stratigraphic inversion: First Break, **16**, no. 9, 311–318.

Duncan, P. M., Gardner, E. B., and Nester, D. C., 1996, Real-time 3-D seismic imaging—A real world case history: 66th Ann. Internat. Mtg., Soc. Expl. Geophys., Expanded Abstracts, 313–315.

Ebrom, D., Li, X., McDonald, J., and Lu, L., 1995, Bin spacing in land 3-D seismic surveys and horizontal resolution in time slices: The Leading Edge, **14**, 37–40.

Embree, P., 1985, Resolutions and rules of thumb: Seismic Field Techniques Workshop

Flentge, D. M., 1996, Method of performing high resolution crossed-away seismic surveys: United States Patent 5 511 039.

Freeland, J. M., and Hogg, J. E., 1990, What does migration do to seismic resolution?: Can. Soc. Expl. Geophys., Recorder, Sept.

Goodway, W. N., and Ragan B., 1995. Focused 3D: Consequences of mid-point scatter and spatial sampling in acquisition design, processing and interpretation: Ann. Mtg., Can. Soc. Expl. Geoph., Expanded Abstracts, 177–178.

Hale, D., 1984, Dip-moveout by Fourier transform: Geophysics, **49**, 741–757.

Harris, C., and Longaker, H. L., 1994, Real time GPS surveying: Recommendations for ensuring a quality survey: Trimble Navigation, internal seminar notes.

Hauser, E. C., 1991, Full waveform recording: An overview of the acquisition, processing, interpretation and practical applications of 3-component seismic for oil and gas exploration: Talisman Energy Inc., internal report.

Head, K., 1998, How could you possibly predict the value of 3-D seismic before you shoot it? Presented at the Geo-Triad Convention.

Hesthammer, J., and Løkkebø, S. M., 1997, Combining seismic surveys to improve data quality: First Break, **14**, no. 2, 103–115.

Hirsche, K., Boerner, S., Kalkomey, C., and Gastaldi, C., 1998, Avoiding pitfalls in geostatistical reservoir characterization: A survival guide: The Leading Edge, **17**, 493–504.

Jack, I., 1997, Time-lapse seismic in reservoir management: 1998 Distinguished Instructor Short Course, Soc. Expl. Geophys.

Jakubowicz, H., 1990, A simple efficient method of dip-moveout correction: Geophys. Prosp., **38**, no. 3, 221–246.

Kallweit, R. S., and Wood, L. C., 1982, The limits of resolution of zero-phase wavelets: Geophysics, **47**, 1035–1046.

Kerekes, A. K., 1998, Shots in the dark . . . : The Leading Edge, **17**, 197–198.

Knapp, R. W., 1991, Fresnel zones in light of broadband data: Geophysics, **56**, 354–359.

Koen, A. D., 1995, Independents step up use of onshore 3D seismic surveys: Oil & Gas Journal, (January 2), 16–20.

Krey, Th. C., 1987, Attenuation of random noise by 2-D and 3-D CDP stacking and Kirchhoff migration: Geophys. Prosp., **35**, 135–147.

La Bella, G., Loinger, E., and Savini, L., 1998, The cross-shooting methodology: design, acquisition, and processing: The Leading Edge, **17**, 1549–1553.

Lansley, R. M., 1994, The question of azimuths: Presented at the SEG Workshop.

Lawton, D. C., 1993. Optimum bin size for converted-wave 3-D asymptotic mapping: CREWES Research Report, **5**, no. 28, 1–14.

Lewis, C., 1997, Seismic attributes for reservoir monitoring: A feasibility study using forward modeling: The Leading Edge, **5**, 459–469.

Ligtendag, M. H. P., 1999, A 3-D data management concept for seismic, workstation support, and interpretation: The Leading Edge, **18**, 330–337.

Lindsey, J. P., 1989, The Fresnel zone and its interpretive significance: The Leading Edge, **8**, 33–39.

Liner, C. L., and Gobeli, R., 1997, 3-D seismic survey design and linear $v(z)$: 67th Ann. Internat. Mtg., Soc. Expl. Geophys., Expanded Abstracts, 43–46.

Lumley, D. E., Behrens, R. A., and Wang, Z., 1997, Assessing the technical risk of a 4-D seismic project: The Leading Edge, **16**, 1287–1291.

Margrave, G. F., 1997, Seismic acquisition parameter considerations for a linear velocity medium: 67th Ann. Internat. Mtg., Soc. Expl. Geophys., Expanded Abstracts, 47–50.

Meinardus, H., and Schleicher, K., 1991, 3-D time-variant dip-moveout by the FK method: 61st Ann. Internat. Mtg., Soc. Expl. Geophys., Expanded Abstracts, 1208–1210.

Neff, W. H., and Rigdon, H. K., 1994, Incorporating structure into 3D seismic survey preplanning: A mid-continent example: Presented at the MESA Technology Conference.

Ongkiehong, L., and Askin, H. J., 1988, Towards the universal acquisition technique: First Break, **6**, no. 2, 46–63.

Padhi, T., and Holley, T. K. 1997, Wide azimuths—why not?: The Leading Edge, **16**, 175–177.

Partyka, G., Gridley, J., and Lopez, J., 1999, Interpretational applications of spectral decomposition in reservoir characterization: The Leading Edge, **18**, 353–360.

Peirce, J. W., Goussev, S. A., Charters, R. A., Abercrombie, H. J., and DePaoli, G. R., 1998, Intrasedimentary magnetization by vertical fluid flow and exotic geochemistry: The Leading Edge, **17**, 89–92.

Pritchett, W. C., 1994, Why waste money with linear sweeps?: The Leading Edge, **13**, 943–948. (With letters in *Signals* section: The Leading Edge, **14**, 66–67.)

Regone, C. J., 1994, Measuring the effect of 3-D coherent noise on seismic data quality: Presented at the 64th Ann. Mtg., Soc. Expl. Geophys.

———1998, Suppression of coherent noise in 3-D seismology: The Leading Edge, **17**, 1584–1589.

Rhodes, J. A. and Peirce, J. W., 1999, MaFIC—Magnetic interpretation in 3-D using a seismic workstation: 69th Ann. Mtg., Soc. Expl. Geophys., Expanded Abstracts, 335–338.

Rozemond, J., 1996: Slip-sweep acquisition: 66th Ann. Mtg., Soc. Expl. Geophys., Expanded Abstracts, 64–67.

Russell, B., 1998, A simple seismic imaging exercise: The Leading Edge, **17**, 885–889.

Schroeder, F. W., and Farrington, R. G., 1998, How fold and bin size can impact data interpretability: The Leading Edge, **17**, 1274–1284.

Schultz, P. S., 1997, The changing economics of 3-D prestack seismic imaging in depth. The Leading Edge, **16**, 471–472.

Schuster, G. T., and Zhou, C., 1996, Is a quasi-Monte Carlo seismic experiment practical?: Presented at the 66th Ann. Internat. Mtg., Soc. Expl. Geophys.

Servodio, R., Picgoli, F., Loinger, E., Pompucci, A. and Schiroli, A., 1997, On the spot 3D—A cost effective seismic tool: Presented at the 59th Ann. Mtg. of the EAGE, Expanded Abstracts.

Sheriff, R. E., 1991, Encyclopedic dictionary of exploration geophysics, 3rd ed.: Soc. Expl. Geophys.

Sheriff, R. E., ed., 1992, Reservoir geophysics: Soc. Expl. Geophys.

Society of Exploration Geophysicists, 1983, Special report on SEG standard exchange formats for positional data: Geophysics, **48**, 488–503.

Steenland, N., 1998, Reflecting on exploration in the North Sea in the 1960s: The Leading Edge, **17**, 479–482.

Trappe, H., and Hellmich, C., 1998, Seismic characterization of Rotliegend reservoirs: From bright spots to stochastic simulation: First Break, **16**, no. 3, 79–87.

Vermeer, G. J. O., 1990, Seismic wavefield sampling: Soc. Expl. Geophys.

———1996, Signals: The Leading Edge, **15**, 10.

———1997, Factors affecting spatial resolution: 67th Ann. Internat. Mtg., Soc. Expl. Geophys., Expanded Abstracts, 27–30.

———1998a, 3-D symmetric sampling: Geophysics, **63**, 1629–1647.

———1998b, 3-D symmetric sampling in theory and practice: The Leading Edge, **17**, 1514–1519.

———1999a, Factors affecting spatial resolution: Geophysics, **64**, 942–953.

———1999b, Converted waves: Properties and 3D survey design: Presented at the 69th Ann. Mtg., Soc. Expl. Geophys., Expanded Abstracts, 645–648.

Wiggins, R. A., Larner, K. L., and Wisecup, R. D., 1976, Residual statics as a general linear inverse problem: Geophysics, **41**, 922–938.

Wilkinson, K., Habiak, R., Siewert, A., and Millington, G., 1998, Seismic data acquisition and processing using measured motion signals on vibrators: 68th Ann. Mtg., Soc. Expl. Geophys., Expanded Abstracts, 144–146.

Wisecup, R. D., 1994, The relationship between 3-D acquisition geometry and 3-D static corrections. 64th Ann. Internat. Mtg., Soc. Expl. Geophys., Expanded Abstracts, 930–933.

Wiskel, B., 1995, Oilman, spare that tree: The Pegg (Feb.).

Wittick, T. R., 1998, Using 3-D seismic data to find new reserves in Quitman Field: The Leading Edge, **17**, 450–456.

Yilmaz, O., 1987, Seismic data processing: Soc. Expl. Geophys.

REFERENCES FOR GENERAL READING

Anderson, N. L., Hills, L. V., and Cederwall, D. A., Eds., 1989, Geophysical atlas of western Canadian hydrocarbon pools: Can. Soc. Expl. Geophys. and Can. Soc. Petr. Geol.

Anderson, R. N., Boulanger, A., He, W., Sun, Y. F., Xu, L., Sibley, D., Austin, J., Woodhams, R., Andre, R., and Rinehart, K., 1995, 4D seismic helps track drainage, pressure compartmentalization: Oil & Gas Journal (March 27), 55–58

Bednar, J. B., 1996, Coarse is coarse of course unless . . . : The Leading Edge, **15**, 763–764.

Canning, A., and Gardner, G. H. F., 1996, Another look at the question of azimuth: The Leading Edge, **15**, 821–823.

Cheadle, S., 1995, Introduction to the 1995 CSEG. 3D wavefield sampling workshop: Ann. Mtg., Can. Soc. Expl. Geophys., Expanded Abstracts, 167–168.

Cooper, N. M., 1995, 3-D design—A systematic approach: Ann. Mtg., Can. Soc. Expl. Geophys., Expanded Abstracts, 171–172.

Dilay, A., and Eastwood, J., 1995, Spectral analysis applied to seismic monitoring of thermal recovery: The Leading Edge, **14**, 1117–1122.

Duncan, P. M., Nester, D. C., Martin, J. A., and Moles, J. R., 1995, 3-D seismic over the Fausse Pointe field: A case history of acquisition in a harsh environment: 65th Ann. Internat. Mtg., Soc. Expl. Geophys., Expanded Abstracts, 966–968.

Durham, L. S. 1995a, Seismic spending can lower costs: AAPG Explorer, **16**, no. 6, 12, 14.

———1995b, By the numbers, 3-D reduces risk: AAPG Explorer, **16**, no. 6, 13.

Eiken, O., and Meldahl, P., 1996, Effects of spatial sampling of marine 3-D seismic data: A case study: The Leading Edge, **15**, 825–830.

Esso Resources Research Company, 1964, 3-D seismic system: In-house pamphlet.

Evans, B. J., 1997. A handbook for seismic data acquisition in exploration: Soc. Expl. Geoph.

Galbraith, J. M., 1995, Seismic processing issues in the design of 3D surveys: Ann. Mtg., Can. Soc. Expl. Geoph., Expanded Abstracts, 175–176.

Goodway, W. N., and Ragan B., 1996, "Mega-Bin" land 3D seismic: Presented at the Ann. Mtg., Can. Soc. Expl. Geoph.

Gordon, I. S., and Voskuyl, J. B., 1995, 3-D planning: Practical design, execution, and economics: Ann. Mtg., Can. Soc. Expl. Geoph., Expanded Abstracts, 173–174.

Gray, S. H., and Etgen J., 1995, 3D prestack migration of overthrust model data: Ann. Mtg., Can. Soc. Expl. Geoph., Expanded Abstracts, 179–180.

Gas Research Institute, 1994, Staggered-line 3D seismic recording: A technical summary of research conducted for Gas Research Institute, the U.S. Department of Energy, and the State of Texas by the Bureau of Economic Geology, The University of Texas at Austin.

Hope, C., Kong, P., and Flentge, D., 1995, 3-D geometry for overthrust areas: The Leading Edge, **14**, 715–719.

Kendall, R. R., and Davis, T. L., 1996, The cost of acquiring shear waves: The Leading Edge, **15**, 943–944.

La Bella, G., Bertelli, L., and Savini, L., 1996, Monti Alpi 3D, A challenging 3D survey in the Appennine range, southern Italy: First Break, **14**, no. 7, 285–294.

Larner, K., Beasley, C. J., and Lynn, W., 1989, In quest of the flank: Geophysics, **54**, 701–717.

Lawton, D. C., 1995, Converted-wave 3-D surveys: Design strategies and pitfalls: Ann. Mtg., Can. Soc. Expl. Geoph., Expanded Abstracts, 69–70.

Lindsey, J. P., 1996, 3-D stack fold: Migration's hidden boost: The Leading Edge, **15**, 847–850.

Liner, C. L., and Gobeli, R., 1996, Bin size and linear $v(z)$: 66th Ann. Internat. Mtg., Soc. Expl. Geophys., Expanded Abstracts, 47–50.

Liner, C. L., and Underwood, W. D., 1999, 3-D seismic survey design for linear $v(z)$ media: Geophysics, **64**, 486–493.

Luzietti, E. A., Moore, D. E., Smith, G. E., Moldoveanu, N., Spradley, M., Brooks, T., and Chang, M., 1995, Innovation and flexibility: Keys to a successful 3-D survey in the transition zone of West Bay field, Louisiana: The Leading Edge, **14**, 763–772.

Meunier, J., and Garotta, R., 1990, Design of land 3-D surveys for stratigraphic purposes: 60th Ann. Mtg., Soc. Expl. Geophys., Expanded Abstracts, 906-910.

Meunier, J. J., Musser, J. A., Corre, P. M., and Johnson, P. C., 1995, Two bottom cable 3-D seismic surveys in Indonesia: Presented at the 65th Ann. Mtg., Soc. Expl. Geophys, 976–979.

Neidell, N. S., 1997, Perceptions in seismic imaging. Part 1: Kirchhoff migration operators in space and offset time, An appreciation: The Leading Edge, **16**, 1005–1006.

O'Connell, J. K., Kohli, M., and Amos, S., 1993, Bullwinkle: A unique 3-D experiment: Geophysics, **58**, 167–176.

Ongkiehong, L., 1988, A changing philosophy in seismic data acquisition: First Break, **6**, no. 9, 281–284.

Oortmann, K. A., and Wood, L. J., 1995, Successful application of 3-D seismic coherency models to predict stratigraphy, offshore eastern Trinidad: 65th Ann. Internat. Mtg., Soc. Expl. Geophys., Expanded Abstracts, 101–103.

Reilly, R. M., 1995, Comparison of circular "strike" and linear "dip" acquisition geometries for salt diapir imaging: The Leading Edge, **14**, 314–322.

Rigo, F., 1995, Overlooked Tunisia reef play may have giant field potential: Oil & Gas Journal, (January 2), 56–61.

Shirley, K., 1995, 3-D now has new mountains to climb: AAPG Explorer, **16**, no. 6, 8, 10 and 16.

Sriram, K. P., and Gilbreth, M., 1995, Intellectual property: Geophysics, **60**, 925–926.

Stark, T. J., 1996, Surface slice generation and interpretation: A review: The Leading Edge, **15**, 818–819.

Stone, D. S., 1994, Designing seismic surveys in two and three dimensions: Geophysical References **5**, Soc. Expl. Geophys.

Thompson, S. D., and Bligh, R. P., 1995, Wytch Farm oilfield, England: Reducing 3-D cycle time and quantifying the benefit for a mature field: 65th Ann. Internat. Mtg., Soc. Expl. Geophys., Expanded Abstracts, 973–975.

Vermeer, G. J. O., 1995, 3D symmetric sampling in land data acquisition: Ann. Mtg., Can. Soc. Expl. Geoph., Expanded Abstracts, 169–170.

Vermeer, G. J. O., den Rooijen, H. P. G. M., and Douma, J., 1995, DMO in arbitrary 3D geometries: 65th Ann. Internat. Mtg., Soc. Expl. Geophys., Expanded Abstracts, 1445–1448.

Walton, G.G., 1972, Three-dimensional seismic method: Geophysics, **37**, 417–430.

Zonati, M., 1996, 3D seismic in the Sirt basin of Libya: First Break, **14**, no. 5, 169–175.

INDEX

3-C 3-D, 177
3-component geophone, 121
3-D symmetric sampling, 79

4-D survey, 177

5-D sampling, 77

85% rule, 41

9-C 3-D, 181

A

achievable resolution, 25
acquisition imprint, 163, 166
active receiver (station), 10
actual resolution, 25
air blast, 111–112, 116
air gun, 124, 175
alias-free sampling, 21
ambient noise, 129, 166
amplitude versus azimuth (AVA) anal, 35
amplitudea variation with offset (AVO), 35, 41, 172, 180
anisotropy, 40, 173
aperture, 11, 25
apron, 11, 47–49
array (or pattern), 8, 77, 129–134
array length, 115, 129
array response, 129–133
aspect ratio, 40–41
azimuth, 39–45

B

basic sampling interval, 88
basic subset, 77–80
bin, 10
bin fractionation, 85–89
bin interval, 20
bin rotation, 169
bin size, 20–28
bird-dog, 148
box, 9, 18, 29–30, 41–43

brick or brick-wall geometry, 83
button patch geometry, 89

C

cable, 121–128, 150
cat push, 74
center-spread acquisition, 77
chair display, 172
charge, 107–113
circular patch, 19, 99
CMP fold, 14–18, 164
common conversion point (CCP), 178–179
common depth point (CDP), 51
common midpoint (CMP), 10, 39, 166
common-offset gather, 79
continuous fold, 79
conversion point, 178
converted wave, 177
correlation, 124
cost model, 76
cost spreadsheet, 69–75
critical reflection, 30, 34–35
crossline, 9
crossline fold, 17
crossline offset, 81
crossline roll, 142
cross-spread, 77–79
curved raypaths, 48, 53

D

deconvolution ("decon"), 111, 124, 155
density, 180
depth migration, 177
depth structure map, 172
diffraction, 12, 26–27, 47–52
digital orthomaps (DOMS), 175
dip moveout (DMO), 162–168
dip versus frequency, 21
direct wave, 34
distributed system, 124–125
double zig-zag geometry, 91–95

drag, 115–117
dual-sensor technique, 176
dwell, 116–118
dynamite, 107
dynamite test, 112

E

economics, 3–4
edge management, 11, 48–51
effort, 114
environment, 8
exclusion area, 135
exploitation, 2

F

far offset, 32–33, 39
Farr diagrams, 155
field QC, 153–156
final survey plan, 135
Flexi-Bin geometry, 85–89
flowchart, 57
fold or fold-of-coverage, 14–19
fold rate, 19
fold striping, 16, 143, 148
fold taper zone, 19
fold-of-illumination, 51, 79
fold versus offset equation, 19, 32
footprint of geometry, 104, 157, 166–167, 173
fracture, 177
frequency, 21
Fresnel zone, 46–47
full-fold 3-D, 77
full-fold area, 11, 19, 50
full swath roll, 147–148

G

geographic information systems (GIS), 175
geometry imprint, 163–165
geophone, 121
geophone array, 129
geophone group, 129
geophysical trespass, 7
geostatistics, 173
global positioning system (GPS), 7, 135–137, 175
ground force, 114–115
ground roll, 111–114, 153, 166
group interval, 40, 60, 121.

H

health, safety, environment (HSE), 126–127
hexagonal binning, 91–96

horizon, 5
horizon slice, 171–172
hydrophone, 121, 175–176

I

illumination area, 178
illumination fold, 52
image area, 50
impedance, 180
inline, 9
inline fold, 16
inline offset, 32
inversion, 180
isotropic, 177

L

largest minimum offset (LMOS), 9
lateral resolution, 24–27
linear moveout (LMO), 155

M

mapping, 171
marsh phone, 121, 175
maximum crossline offset, 77
maximum inline offset, 77
maximum recorded offset, 11
maximum unaliased frequency, 21
Mega-Bin, 91–96
midpoint, 10
midpoint density, 20
midpoint scatter, 40
migration, 25–26
migration and random sampling, 167
migration aperture, 11
migration apron, 10
migration noise, 167
minimal data set, 79
move-up, 115
multiple, 35
multiple suppression, 35, 41, 166
mute function, 31

N

narrow azimuth geometry, 40–41
natural bin, 20
near-offset trace, 60–61
NMO discrimination, 35
NMO stretch, 35
no permit area, 99
nominal fold, 99
nonorthogonal geometry, 83
normal moveout (NMO), 159

O

ocean-bottom cable technique (OBC), 176
offset, 9
offset distribution, 39
orthogonal geometry, 14, 81

P

pad time, 114
parallel geometry, 81
patch, 10
peak force, 117–118
permitting, 6, 99, 150
phase, 117
porosity, 4, 180
prestack depth migration, 177
prestack time migration, 176
processing, 157

Q

QC, 147, 153–156, 157
quasi-random, 99

R

radial, 97
random geometry, 96–99
ray bending, 22
ray-trace modeling, 51
Rayleigh criterion, 28
receiver line, 9
reciprocity, 166
record length, 52
recorders, 123
reflection statics, 161
refraction statics, 158
reservoir management, 177
reservoir monitoring, 181
roll-on/off, 141–143
rule of thumb—fold, 14
rule of thumb—fold taper, 19
rule of thumb—migration apron, 48
rule of thumb—selection of bin size, 25
rule of thumb—X_{min}, 31
rule of thumb—X_{max}, 34

S

safety, 7–8, 135, 151–152
scattering angle, 11, 25, 47–48, 52
scouting, 5
script file, 61, 138
SEG-P1 format, 135
SEG-Y format, 158
semblance, 159
SH wave, 178
shear wave, 178
shooting strategy, 145–146
shot, 107
shot density, 10, 15
shot hole, 111–112
signal-to-noise ratio (S/N), 10, 14, 20, 166
similarity tests, 116
skids, 149
source, 107
source density, 10, 15
source line (shotline), 9
source-receiver pair, 10, 39
spatial continuity, 79–80, 154
spatial frequency, 26.
spider diagram, 40, 44
spread, 121
SPS format, 138–141
stack, 166
stack array, 131
standardized spreadsheets, 62–69
star shooting, 96–99
statics coupling, 161–162
statics, 31–32, 104, 146, 158, 161–162
straight line, 16, 97, 162
streamer acquisition, 81, 175
sub-bin, 85
super bin, 10, 104, 159
swath, 10, 80–81, 143–145, 147–148, 162
S-wave, 113, 178–179
sweep, 114
sweep length, 114
sweep rate, 117
symmetric sampling, 30, 79, 134

T

takeout, 121, 150
target depth, 25, 34, 49, 111
target size, 21
telemetry system, 121, 126
template, 10, 41–42, 138, 141–147
TIFF file, 153, 175
time slice, 21, 155, 166–167, 172
time structure map, 171
time-lapse survey, 177
total nominal fold, 18
transition zone, 7–8, 175–176

U

uncorrelated record, 118
undershooting, 47, 119

unit cell, 9
UTM, 137

V

vari-sweep, 117
velocity, 159
velocity model, 177
vertical resolution, 27
vibrator, 7, 91, 113–118, 131, 146
vibroseis, 107

W

wavelength, 21–28, 46–48
wavelet, 27–28, 47, 123, 180
wavenumber, 25–26

weathering layer, 112
well tie, 180
wide-azimuth geometry, 41–44
Widess criterion, 27

X

X_{max}, 11, 32–36, 45–46, 60
X_{min}, 11, 29–32, 60
X_{mute}, 11, 32, 41, 45–46

Z

zero offset, 46, 162–163
zig-zag geometry, 91–96
zipper design, 146
zone of influence, 47